ANATOMY AND PHYSIOLOGY
An Illustrated Guide

 **Marshall Cavendish**
Reference
New York

Marshall Cavendish

99 White Plains Road

Tarrytown, NY 10591–9001

www.marshallcavendish.us

Library of Congress Cataloging-in-Publication Data

Anatomy and physiology : an illustrated guide.
    p. cm. -- (Marshall Cavendish reference)
  Includes bibliographical references and index.
  ISBN 978-0-7614-7881-2 (alk. paper)
  1. Anatomy--Atlases. 2. Physiology--Atlases. I. Marshall Cavendish Corporation.
  QM25.A4887 2010
  612--dc22

                                      2009002177

Printed in Malaysia
13 12 11 10 09   1 2 3 4 5

MARSHALL CAVENDISH
Publisher: Paul Bernabeo
Production Manager: Mike Esposito

THE BROWN REFERENCE GROUP PLC
Managing Editor: Tim Harris
Subeditors: Jolyon Goddard, Paul Thompson
Designer: Bob Burroughs
Picture Researcher: Laila Torsun
Indexer: Kay Ollerenshaw
Design Manager: David Poole
Editorial Director: Lindsey Lowe

PHOTOGRAPHIC CREDITS
Cover illustration is a composite of anatomical illustrations found within this work.

ARS: 56b; **Corbis:** Hall Beral 142; Lester V. Bergman 55, Bettmann 56t, 85, Matthias Kulka 82; Louie Psihoyos; 174; Dennis Wilson; 29; **FLPA:** Silvestris Fotoservice 136; Reg Morrison/Auscape/Minden Pictures 158; D. P. Wilson 148; **OSF:** Carolina Bio Supply Co. 26; Phototake 23, 127, 134; **NaturePL:** Adrian Davies 65 Bruce Davidson, 159; Jeff Foott, 183; Dietmar Nill 146; Constantinos Petrinos 167; Jeff Rotman 176, Jose B. Ruiz 163; **NHPA:** Peter Parks, 133; **Photodisc:** 71, 129; **Photos.com:** 97, 132, 151, 152–153; **Rex:** Voisin/Phanie 88, 91; **SPL:** Biophoto Associates 148; David Becker 125; Eye of Science 33, 93, 168; Steve Gschmeisser 105, 113, 114; Dr Elena Kiseleva; 22; Bsip Lecaque 84; David McCarthy 77; Peter Menzel 90; Susumu Nishinaga 117; **Philippe Plailly:** 40; **VVG:** 31; **Still Pictures:** Kelvin Aitkin 181; John Cancalosi 172, Dani 63; Andrew Davies 178; **Manfred Kage:** 17, 162; Steve Kaufman 143; Dr. Michael Klein, 182; **Alfred Pasieka:** 18; **Ed Reschke:** 7, 16, 99, 103; Roland Seitre 169; Howard Sochurek/The Medical File 170; SIU 188; Dr D. Spector: 11.

Artworks: The Art Agency, Mick Loates, Michael Woods.

Contents

CONSULTANTS AND CONTRIBUTORS

CONSULTANTS

- Barbara J. Abraham, PhD, Interim Chair, Department of Biological Sciences, Hampton University, Hampton, VA. • Glen Alm, MSc, Mushroom Research Program, University of Toronto, Ontario, Canada. • Roger Avery, PhD, former Senior Lecturer in Zoology, Bristol University, England. • Amy-Jane Beer, PhD, Director of natural history consultancy Origin Natural Science. • Deborah Bodolus, PhD, Department of Biological Sciences, East Stroudsburg University, PA. • Allan J. Bornstein, PhD, Department of Biology, Southeast Missouri State University, Cape Girardeau, MO. • Erica Bower, PhD, consultant to Royal Botanic Gardens, Kew, England. • John A. Cline, PhD, Assistant Professor in Tree Fruit Physiology, Department of Plant Agriculture, University of Guelph, Ontario, Canada. • Trevor Day, marine scientist and visiting lecturer, University of Bath, England. • John Friel, PhD, Curator of Fishes, Amphibians, and Reptiles, Cornell University Museum of Vertebrates, Research Associate, Department of Ecology and Evolutionary Biology, Cornell University, NY. • Valerius Geist, PhD, Professor Emeritus of Environmental Science, University of Calgary, Alberta, Canada. • John L. Gittleman, PhD, Scientific Fellow of The Zoological Society of London and Professor of Biology, University of Virginia, Charlottesville, VA. • Tom Jenner, PhD, teacher, Academia Britanica Cuscatleca, El Salvador. • Bill Kleindl, MSc, aquatic ecologist. • Thomas H. Kunz, PhD, Director, Center for Ecoology and Conservation Biology, Boston University, MA. • Alan C. Leonard, PhD, Professor of Biological Sciences, Florida Institute of Technology, Melbourne, FL. • Sally-Anne Mahoney, PhD, neuroscience researcher, Bristol University, England. • Chris Mattison, herpetologist and author, Sheffield, England. • Andrew S. Methven, PhD, Professor and Chair, Department of Biological Sciences, Eastern Illinois University, Charleston, IL. • Graham Mitchell, PhD, Malaria Laboratory, GKT School of Medicine, Guy's Hospital, London, England. • Richard J. Mooi, PhD, Curator of Echinoderms, California Academy of Sciences, San Francisco, CA. • Ray Perrins, PhD, former neuroscience researcher, Mount Sinai Medical Center, New York. • David Spooner, PhD, Professor of Horticulture, University of Wisconsin, Madison, WI. • Adrian Seymour, PhD. Senior Forest Scientist, Operation Wallacea Indonesia Program. • John Stewart, BSc, researcher, Natural History Museum, London, England. • Erik Terdal, PhD, Associate Professor of Biology, Northeastern State University, Broken Arrow, OK. • Philip J. Whitfield, PhD, Professor, School of Health and Life Sciences, Kings College, University of London.

CONTRIBUTORS

- Amy-Jane Beer, PhD, Director of natural history consultancy Origin Natural Science. • Erin L. Dolan, PhD, Fralin Biotechnology Center, Blacksburg, VA. • Bridget Giles, BA, natural history writer, London, England. • Natalie Goldstein, natural history writer, New York. • Christer Hogstrand, PhD, Adjunct Associate Professor, University of Miami, FL, and lecturer at King's College, University of London. • James Martin, BSc, natural history writer, London, England. • Graham Mitchell, PhD, Malaria Laboratory, GKT School of Medicine, Guy's Hospital, London, England. • Ray Perrins, PhD, former neuroscience researcher, Mount Sinai Medical Center, New York.

Foreword

The world runs on the successful interlocking of systems. Electricity, transportation, waste management, and air quality all depend on systems working together to run cities and countries. Anyone who wants to understand how a civilization survives sooner or later comes to grips with the integration of systems.

It is no different when it comes to the study of living things. The analysis of the inner workings of organisms, be they plants or animals, requires an appreciation of anatomy and physiology—the forms and functions of life. These scientific disciplines comprise the keys to biological knowledge, helping to develop a broad view of how organisms, no matter how distantly related, cope with the dictates of reproduction, feeding, excretion, gas exchange, locomotion, and a host of other activities that characterize life on Earth. The old adage that "form follows function" is no better expressed than in the diversity of physiological responses that animals and plants have evolved as adaptations to environmental factors. What is striking about these adaptations is not just the fascinating array of solutions that have come and gone over time, but the remarkable number of times that organisms hit upon similar answers to similar situations, even though the organisms involved might be very distant from one other on the tree of life. Life forms as different as grasses, sponges, and birds all need to meet the requirements of producing the gametes used in sexual reproduction. Skeletal systems support sea urchins, corals, and cats. The challenges of gas exchange and respiration in water and in air have been met by fish, beetles, and the leaves of all sorts of plants.

Anatomy and Physiology: An Illustrated Guide introduces the world of physiological systems in a brilliantly illustrated and comprehensive way. Like street maps of cities, the sections of this work lead through the twists and turns of life's inventiveness, telling the stories of how physiological systems operate. With carefully researched text and detailed drawings reviewed by experts in the field, these articles make it easy to follow how each system works and contributes to the lives of organisms. The latest concepts in physiological science, including recent and remarkable advances in genetics and organic development, are clearly laid out for the student and expert alike. As a resource for coursework or simply for taking an informative tour through the organismal neighborhood, *Anatomy and Physiology* will be an indispensable companion.

Richard Mooi

Richard Mooi is Curator of Echinoderms at the California Academy of Sciences, San Francisco

The articles on biological systems included in this work are also available by subscription online from Marshall Cavendish Digital at www.marshallcavendishdigital.com as part of a larger encyclopedic work, *Animal and Plant Anatomy*, which also contains more than 80 additional articles on the anatomy of particular organisms.

Cell biology and genetics

All living things have two goals: to survive and to reproduce. The smallest unit of life is the cell, and cells also have these goals. The sciences of cell biology and genetics examine how cells survive and reproduce.

Living things are grouped into two major categories, called prokaryotes and eukaryotes, depending on the basic structure of their cells. Bacteria are prokaryotes, and all other living things, including plants, animals, fungi, and protists, are eukaryotes.

Single-celled organisms, including bacteria and simple eukaryotes such as amoebas, ciliates, and various algae, are interesting to biologists because they manage to accomplish all the tasks required to sustain life using one multipurpose cell. In multicelled organisms, different cells tend to be specialized for different roles, such as acquiring food, using energy, eliminating waste, avoiding harm, and producing offspring. The shape and structure of cells depend on what functions a cell has to accomplish. How cell structures are made is defined by genes, which contain instructions for making proteins. Proteins are immensely diverse molecules that contribute to both the structure and the functioning of the cell. A group of proteins known as enzymes act as enablers, or catalysts, in a vast number of important chemical reactions. Without them, the processes of life would grind to a halt.

What are cells made of?

Cells are made of four major types of macromolecules, or large molecules: proteins, carbohydrates, fats (also known as

▼ An animal cell contains many organelles that together enable it to eat, grow, reproduce, and fight off invaders.

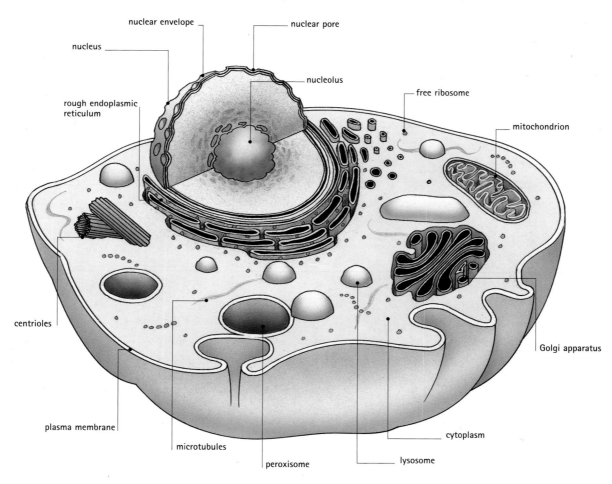

nuclear envelope

nuclear pore

nucleus

nucleolus

rough endoplasmic reticulum

free ribosome

mitochondrion

centrioles

Golgi apparatus

plasma membrane

cytoplasm

microtubules

lysosome

peroxisome

lipids), and nucleic acids. In general, proteins are the active components of the cell, performing most of the functions essential to life. Carbohydrates embellish the proteins, modifying and fine-tuning them to do specific jobs and helping provide structure to the cell itself. Lipids, which are more malleable than proteins or carbohydrates, make up the outer layer of cells. Nucleic acids carry information for manufacturing all of the molecules within cells. Deoxyribonucleic acid (DNA), which contains the sugar deoxyribose, stores this information, like a recipe book for the cell. Ribonucleic acid (RNA) contains the sugar ribose. When a cell needs to make a particular protein, RNA accesses the necessary information coded in the DNA and translates it into a form recognized by the protein-building components of the cell.

The beginnings of life

How did life get started? Scientists have studied the chemical conditions they think existed on Earth many millions of years ago and found evidence that RNAs may have been present. This discovery caused some excitement because RNA can store genetic information and it can act as an enzyme, called a ribozyme. Enzymes, most of which are proteins, are molecules that enable chemical reactions to occur. Chemical reactions underlie almost all processes in living organisms, including those necessary for survival and reproduction. Thus RNA may have been the first molecule that not only was able to perform the chemical reactions necessary for life (survival), but also could pass on genetic information (reproduction).

While RNA may be the molecule with which life began, it is still unclear how the next step took place. How did a chemical reaction become a living, biological entity? One can imagine that molecules acting together can accomplish more tasks than those acting alone, and the simplest life-forms must have evolved from groups of

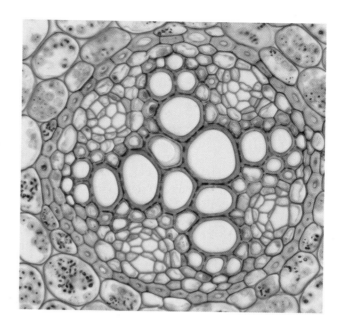

▲ This cross section of a buttercup root shows the different types of cells stained for clarity. The xylem cells (red) are used for transporting water, and the phloem cells (green) are used for transporting nutrients. Surrounding these are the pericycle (yellow), involved in the formation of lateral roots; and the cortex (large green cells), used for storage of nutrients.

molecules working together. Very simple cells, such as bacteria, are enclosed bundles of chemical reactions that can use energy to acquire food, eliminate waste, move, and reproduce. The reactants are enclosed within a malleable balloonlike container—the cell membrane—and a tough structure that provides protection and physical support: the cell wall. More complex life-forms evolved when groups of cells began to work together.

Cells and their functions

Cells in multicellular organisms specialize to become different in structure and function. For example, animal nerve cells, called neurons, transmit information gathered from all around the body and from the outside world to the brain, and then trigger an appropriate response. Specialized neurons receive different kinds of information from various sense organs: visual information through eyes or light-sensitive cells; sound information through ears; touch information from whiskers or skin; and chemical information such as taste and smell. Chemical receptors are located in the mouth and nose of most vertebrates but may be mounted on antennae or scattered all over the body surface in other animals. Some animals have additional senses, for example for detecting vibrations, electrical signals, or magnetism. However, regardless of its source, all sensory information is ultimately processed by some part of the nervous system, and neurons are specialized to do the job as quickly and efficiently as possible.

Muscle cells

Differentiation is the process by which cells become specialized. Neurons are just one example, and muscle cells are another. Muscle cells, or myocytes, make proteins that allow the cells to contract. When muscle cells work together, they can generate movement on a much larger scale—that of the whole muscle or whole animal. Humans and other vertebrates have three types of muscle cells: skeletal, smooth, and cardiac. Skeletal muscles are attached to the skeleton and move the torso and limbs. As skeletal muscle cells form, they align to create long linear fibers, hence their alternative name: striated (or striped) muscle. Skeletal muscle cells also fuse to create larger

---

IN FOCUS

How cells eat

All cells need to harvest macromolecules to provide the building blocks of their own proteins, carbohydrates, fats, and nucleic acids. Cells also need energy to assemble and disassemble these molecules. Autotrophs, or "self-feeders," can create their own macromolecules from inorganic raw materials, such as carbon dioxide, water, and light energy. Plants and bacteria that use photosynthesis to construct sugar and other molecules are autotrophs. Animals and fungi are heterotrophs, which obtain macromolecules and energy by eating or decomposing the tissues or cells of other organisms. All heterotrophs are ultimately dependent on autotrophs for food, and the vast majority also depend on oxygen, a byproduct of photosynthesis.

▶ *In phagocytosis, large solid particles are enclosed by folds of the plasma membrane. In pinocytosis, the membrane folds inward, beneath the molecule, enclosing dissolved material.*

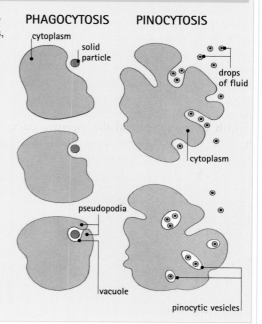

cells with many nuclei, which contract in one dimension. Smooth muscle cells are found in the walls of the stomach, intestines, and blood vessels. These cells are smaller and less organized than skeletal muscle cells and contract in waves. Cardiac muscle cells assemble to form the heart, which contracts repeatedly and without tiring to pump blood around an animal's body.

The sex cells

Sex cells, or gametes, are specialized to create offspring. Male gametes, called spermatozoa or sperm, have a tail called a flagellum (plural, flagella) that helps them move to find an egg for fertilization. Female gametes, called eggs or ova (singular, ovum), are typically large and carry nutrients and energy to sustain the zygote, or fertilized egg, at the beginning of development. Both eggs and sperm are haploid, carrying only half the DNA normally needed by a new individual. When each contributes its DNA to the zygote, the zygote is then diploid, having two copies of each chromosome. If sex cells were diploid, the next generation would have four sets of instructions (tetraploid), the next would have eight copies (octoploid), and so on. Meiosis is the process of cell division by which the number of DNA-containing structures, called chromosomes, is halved to produce haploid gametes.

Guard cells

Other cells guard against damage. Epithelial cells cover all surfaces in the human body, holding organs and tissues together and protecting them from invaders like bacteria and viruses. Epithelial cells also cover internal surfaces such as the respiratory tract, from the nostrils to the lungs; and the digestive tract, from the mouth to the stomach, intestines, and anus. Skeletal tissues, such as bones and cartilage, guard against damage by giving the body structure and protecting internal organs. For example, the rib cage protects the heart, lungs, and liver; and the skull and vertebrae protect the brain and spinal cord.

Plant cells

Plants also have specialized cells. Parenchyma cells in leaves contain chloroplasts that gather light energy and use it to manufacture the sugar glucose from carbon dioxide and

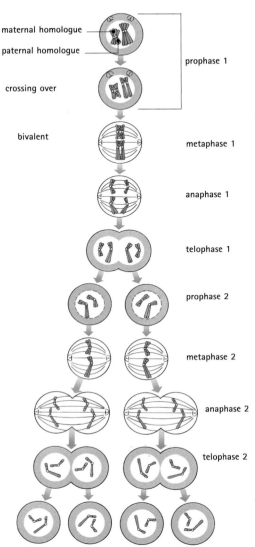

maternal homologue

paternal homologue

prophase 1

crossing over

bivalent

metaphase 1

anaphase 1

telophase 1

prophase 2

metaphase 2

anaphase 2

telophase 2

four genetically distinct haploid cells (gametes)

◄ MEIOSIS

Each chromosome is made up of a pairing of a paternal homologue (from the father) and a maternal homologue (from the mother). At prophase 1, the chromosomes duplicate, forming X-shaped chromosomes. During a process called crossing over, the chromosomes that form these pairs exchange some genetic material. Telophase 1 produces two new cells containing half the full complement of chromosomes: that is, they are haploid. In prophase 2 onward there is no duplication of DNA. The subsequent phases result in the production of two sibling chromatids, which separate to produce four distinct haploid offspring cells. When a male gamete (sperm) fertilizes (fuses with) a female gamete (egg or ova) the resulting cell, or zygote, has two sets of chromosomes; so, it is diploid.

water using a process called photosynthesis. Parenchyma cells in the stem and roots store starch. Collenchyma and sclerenchyma cells help provide structural support. Collenchyma cells are stiff but flexible, and are found in younger plants because they give the plant structure without restricting its growth. Sclerenchyma cells are less flexible and are found in older plants. Vascular plants also contain two sorts of conducting tissues: xylem, which carries water from the roots to the rest of the plant; and phloem, which moves food molecules throughout the plant.

Cell external anatomy

Cells have specialized structures depending on their functions. All cells are surrounded by a cell membrane that acts like a sack to contain the cell contents. The cell membrane also selects which molecules can move into and out of the cell, keeping out toxic materials and bringing in nutrients and beneficial molecules. In addition to the cell membrane, plant cells have a rigid cell wall, which protects against damage and maintains the plant's structure.

A cell's shape is tailored to its function. For example, the chains of neurons that receive touch information in human fingers are very long because they have to reach from the spinal cord to the fingertips. The neurons in your ear, in contrast, have hairlike protrusions that move in response to sound vibrations. The neurons that sense pressure on your skin are buried deep; those that sense pain spread fine spidery endings very close to the surface.

The external surface of some cells can be studded with a huge variety of structures. Single-celled organisms in particular may be covered with cilia or other tiny hairlike

▼ PLASMA MEMBRANE
All cells are surrounded by a plasma membrane, which allows some molecules to pass into and out of the cell but denies access to others. The plasma membrane is thought to consist of two layers of phospholipid molecules with globular proteins embedded among them, arranged in a random mosaic pattern.

CLOSE-UP

The passage of molecules

Cell membranes select which molecules are allowed to move into and out of cells by using specialized pores called channels, or proteins called receptors. Channels allow only certain molecules, like sodium or potassium ions, across the membrane. Channels can be gated, so that molecules are allowed through only under certain conditions, such as when electrical charges change or particular hormones interact with the cell. Receptor and transporter proteins sit on or in the membrane, ready to bind a particular kind of molecule, known as the substrate. The substrate fits the receptor protein like a key in a lock. When a substrate binds its receptor, the receptor signals the inside of the cell to respond or to bring the substrate inside the cell. Transporter proteins are like receptors, but receptors only bind molecules whereas transporters move molecules across the membrane into or out of the cell.

structures, plates of mineral or organic armor, or sheaths of gelatinous ooze, which are manufactured by the cell itself. Most cells bear thousands of different receptors that enable signaling chemicals such as hormones to react with cells, and others that allow cells to interact with one another. Cells also have channels and transport proteins that allow some molecules to cross the membrane but keep others either in or out.

In multicellular organisms, cells are often surrounded by an extracellular matrix (ECM), a web of material that holds cells in place and stores materials for later use. The ECM is especially important during development, because it provides a pathway along which cells locate their destinations. The ECM degenerates with age, leading to less elastic, sagging tissue.

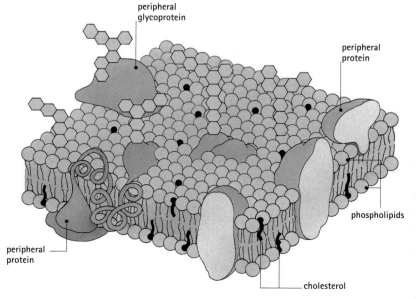

peripheral glycoprotein

peripheral protein

peripheral protein

phospholipids

cholesterol

Cell internal anatomy

Eukaryotes, including plants, animals, and fungi, have a cell membrane, and plants and fungi also have a cell wall. One of the main differences between prokaryote and eukaryote cells is that chemical reactions in the latter are often contained within small compartments, called organelles, or "little organs." Examples of organelles include mitochondria (singular, mitochondrion), which harness energy from food; and nuclei (singular, nucleus), which hold genetic information and control how this information is used. Other organelles include lysosomes, endoplasmic reticulum, and the Golgi apparatus, all of which are essential for the construction of proteins. The first appearance of organelles marked another major step in the evolution of life. The "endosymbiont" theory proposes that eukaryote cells first formed when one prokaryote cell engulfed another. Instead of being digested as food, the engulfed cell continued living within the host, for which some of its functions served a useful purpose. Thus the engulfed cell began acting as an organelle. The fact that mitochondria are enclosed by a double membrane and contain their own DNA supports the idea that they have evolved from independent cells.

The work of cells

One of the most important jobs of a cell is to manufacture proteins. Many proteins are made as part of routine maintenance, just to keep the cell alive and healthy, but others are made in response to some outside influence. For example, when an animal like a dog or human eats, its digestive system breaks down the food into basic component molecules, which can be taken up by intestine cells and then transported all over the body. One of the most important of these is the sugar glucose. Receptors in an

▼ CELL STRUCTURES
This electron microscope image of a cell shows several internal structures. The curved strands are the rough endoplasmic reticulum, the dots on these curves are ribosomes, and the large purple structures are mitochondria.

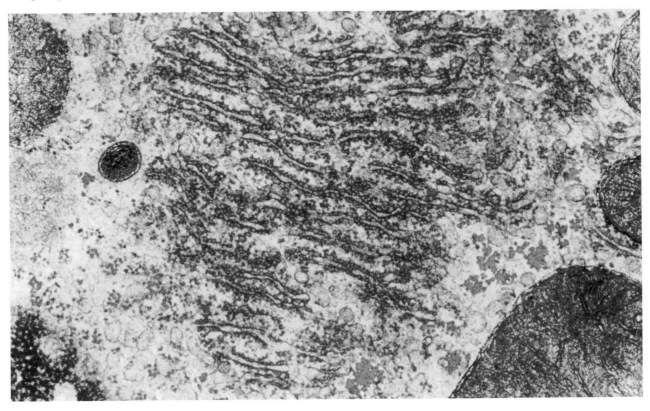

COMPARE the genetic material of a eukaryotic cell with that of prokaryotic cells such as *BACTERIA*. In the eukaryotic cell, the DNA is arranged within a nucleus, but a prokaryotic cell does not have a distinct nucleus.

organ called the pancreas detect increased levels of glucose in the circulation and trigger its cells to manufacture and release a protein called insulin. Insulin is a hormone. Having been released into the blood, it is carried around the body, binding to specialized receptors on many other cells and encouraging them to take up glucose using a protein called a glucose transporter. Glucose can then be used as an energy source for all of the cell's functions. This may seem like a simple pathway, but it is actually a complex set of steps involving all parts of the cell.

Nucleus

Cell nuclei are the storage sites for chromosomes in all eukaryotic cells. The gene needed to make insulin, for example, is stored in the nucleus of every cell in the animal's body, but it is active only in those differentiated into pancreas cells. The nucleus has a membrane envelope that protects the chromosomes from damage and controls which genes are used to make proteins. Molecules called messenger RNAs (mRNAs) are made from the genes coded on particular strands of DNA.

Ribosomes

Ribosomes are tiny granular particles that assemble proteins. There are vast numbers of ribosomes in a cell. Ribosomes read mRNA transported out of the nucleus and use it as a template for building the specific protein required, for example, insulin. Some ribosomes are free-floating in cells. The proteins they make will also be free-floating in the cytoplasm. Other ribosomes are attached to a large network of membranes called the endoplasmic reticulum (ER).

Endoplasmic reticulum

ER that is covered with ribosomes is called rough endoplasmic reticulum. The ribosomes use mRNAs to manufacture proteins that will be built into membranes or packaged for storage or export in little bubbles called vesicles. When a cell needs a new membrane protein, such as a receptor in the cell membrane, the nuclear membrane, or the membrane of any other organelle, its ribosomes will weave the protein back and forth in the ER membrane as they assemble it. Small pieces of ER containing the new protein can be broken off, transported to other locations in

▶ BACTERIAL RIBOSOME
Electron microscopes have been used to discover the position of many proteins in the structure of a bacterial ribosome. The ribosome is made of two subunits that together use a form of RNA called messenger RNA to assemble proteins. This illustration shows the subunits, and the way they interlock, from two different angles.

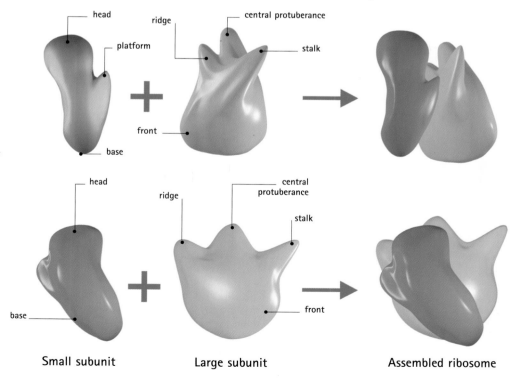

Small subunit   Large subunit   Assembled ribosome

the cell, and patched into the membrane where they are needed. When a protein is to be released outside the cell, rather than weaving the new protein into the membrane, the ribosome pushes it into the ER cavity, or lumen. Then small, bubblelike vesicles filled with the protein are budded off the main ER and can be moved to other parts of the cell. If the protein is to be exported, the vesicle moves to the cell membrane and fuses with it, expelling the contents outside the cell.

ER that does not contain ribosomes is known as smooth endoplasmic reticulum (SER). SER is involved in synthesizing steroid hormones like estrogen and testosterone and also helps detoxify drugs and other poisons.

Golgi apparatus

Most proteins synthesized on the ER need some kind of further processing before they can be useful. All this processing occurs in a structure called the Golgi apparatus, which looks like a stack of pancakes. The Golgi has two sides: cis, where proteins are received from the ER; and trans, from which processed proteins are dispatched to other parts of the cell. Proteins passing through the Golgi are trimmed to their correct size, assembled into groups, folded into intricate shapes, and decorated with sugars. These sugars have many functions—they help the protein fold correctly and allow it to be recognized by other cells, organelles, or molecules, for example.

▼ GOLGI APPARATUS AND ENDOPLASMIC RETICULUM
These two membranous organelles are involved in the production of many molecules necessary for the functioning of the cell. The manufacture of lysosomal proteins, for example, begins in the endoplasmic reticulum with the production of proteins with mannose sugars. The proteins then pass to the Golgi apparatus where phosphates are added, thus forming lysosomal proteins. The proteins are then transported to a lysosome where they break down waste material.

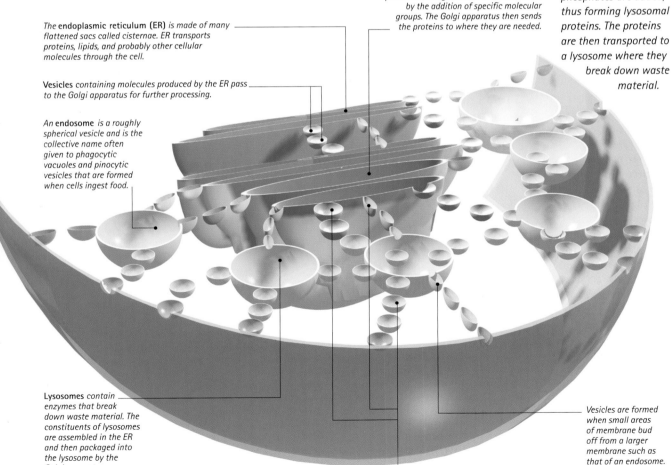

The Golgi apparatus is also made of flattened sacs called cisternae. There, proteins arriving from the ER are modified by the addition of specific molecular groups. The Golgi apparatus then sends the proteins to where they are needed.

The endoplasmic reticulum (ER) is made of many flattened sacs called cisternae. ER transports proteins, lipids, and probably other cellular molecules through the cell.

Vesicles containing molecules produced by the ER pass to the Golgi apparatus for further processing.

An endosome is a roughly spherical vesicle and is the collective name often given to phagocytic vacuoles and pinocytic vesicles that are formed when cells ingest food.

Lysosomes contain enzymes that break down waste material. The constituents of lysosomes are assembled in the ER and then packaged into the lysosome by the Golgi apparatus.

Vesicles are formed when small areas of membrane bud off from a larger membrane such as that of an endosome.

Membrane-bound vesicles shuttle between the cell wall, lysosomes, endosomes, Golgi apparatus, and ER. These vesicles contain many substances necessary for the functioning of the cell, such as proteins, lipids, nutrients, and digestive enzymes.

Helpful chemicals: Vitamins and antioxidants

Vitamins are molecules that perform vital functions in the normal activity of cells, smoothing and enhancing various metabolic reactions. Vitamins B and C, for example, are needed to make cofactors, molecules that activate various enzymes. Vitamin A is converted into retinal, a molecule that helps nerve cells in the eye detect light. Different organisms are capable of synthesizing different vitamins, but often there are others that have to be taken in ready-made as part of the diet. Most animals can manufacture their own vitamin C, but primates, including humans, can get it only from their food; that is why fruit is an important part of the human diet. Vitamin deficiencies can lead to severe problems.

Many cell processes involve the movement of electrons from one molecule to another. Molecules that have lost electrons are said to be oxidized. Oxidation generates reactive molecules called free radicals, which can harm cells by damaging DNA and other important molecules. Free radicals contribute to the effects of aging, and play a role in health problems as diverse as heart disease, types of cancer, arthritis, and Alzheimer's disease. Antioxidants are chemicals that normally neutralize free radicals. Many vitamins, including vitamins C and E, are important antioxidants. The plant pigment beta-carotene, found in tomatoes and carrots, is another.

Mitochondria

Almost every process in the cell, from sending signals to making proteins, requires energy. Mitochondria (singular, mitochondrion) are the organelles that harness the energy from glucose and convert it into a form that can be used by other parts of the cell. Once glucose has been transported into the cell, it is quickly converted to a related molecule called pyruvate, or pyruvic acid, which cannot easily move back out of the cell. Pyruvate is then broken down in the mitochondria. Pyruvate is used to manufacture a chemical called adenosine triphosphate (ATP). ATP is the universal energy storage molecule and can be used by all organisms to sustain life.

▶ **MITOCHONDRION**
Mitochondria use fats and sugars in a complex series of chemical reactions to create a molecule called adenosine triphosphate (ATP), which is used as an energy store. Fats and sugars enter the mitochondrion and undergo processes called beta oxidation and the citric acid cycle. Protein complexes (labeled I to V) on the inner membrane together perform a further series of processes collectively called the electron transport chain. This results in the production of ATP, which is released into the cytoplasm of the cell.

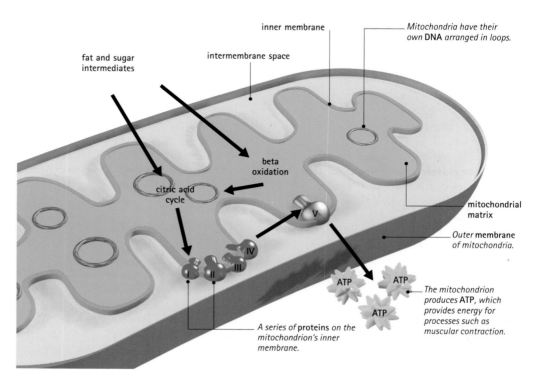

inner membrane

intermembrane space

fat and sugar intermediates

Mitochondria have their own **DNA** arranged in loops.

beta oxidation

citric acid cycle

V

IV

III

II

I

mitochondrial matrix

Outer **membrane** *of mitochondria.*

ATP

ATP

ATP

A series of **proteins** *on the mitochondrion's inner membrane.*

The mitochondrion produces **ATP**, *which provides energy for processes such as muscular contraction.*

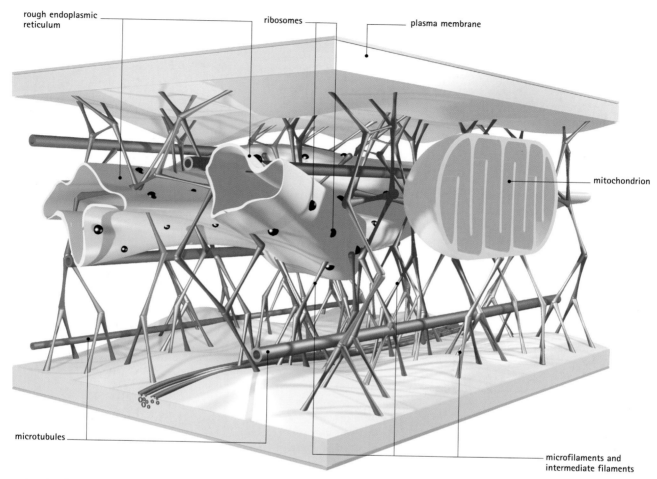

rough endoplasmic reticulum

ribosomes

plasma membrane

mitochondrion

microtubules

microfilaments and intermediate filaments

A mitochondrion is surrounded by two layers of membranes, which create two separate compartments. Pyruvate is broken down in the innermost compartment of the mito-chondrion, inside the inner membrane. Then several negatively charged particles (electrons) are extracted from it. This process releases energy to pump positively charged particles (protons), from between the two mitochondrial membranes to the innermost compartment. The protons then flow back between the two layers, providing energy to generate ATP. This whole process, which requires glucose and oxygen, is called cellular respiration.

Lysosomes and peroxisomes

Making proteins and using energy generate waste. Lysosomes and peroxisomes are structures inside the cell that use lysosomal protein produced in the endoplasmic reticulum and Golgi apparatus to gather waste, break it down, and neutralize potentially damaging molecules, such as hydrogen peroxide. The neutralized waste can then be eliminated from the cell through a process called exocytosis.

Cytoplasm and cytosol

All cell parts are suspended in a watery medium called cytosol, which bathes the organelles and helps the cell maintain its shape. Cytoplasm is made up of the cytosol and all the other cell contents, including organelles. In order for all cell processes to function properly it is essential that the cytoplasm contains the right mixture of chemicals to create optimum acidity (pH) and salt, glucose, and water content. Chemical reactions in the cytoplasm also affect the temperature of a cell.

▲ CYTOSKELETON

The cytoskeleton is a network of protein microfilaments, microtubules, and intermediate filaments that extends throughout the cell's cytoplasm. The cytoskeleton provides structure for the cell and its organelles and is involved in cell movement.

Plant vacuoles

Plant cells have specialized organelles called vacuoles that store water, nutrients, and other molecules. The contents of a vacuole create pressure called turgor, which helps give rigidity to the cell. As long as its main vacuole stays full of water or food, a plant cell stays plump and inflated. Vacuoles in certain cells have more specific functions: those in flower cells, for example, store pigments that give petals their bright colors.

Chloroplasts

Plant cells have organelles called chloroplasts that use carbon dioxide and water to make sugars using energy from light. This process is called photosynthesis. Like mitochondria, chloroplasts have their own DNA and are thought to have originated when one cell engulfed another, leading to a symbiotic relationship. Chloroplasts contain the pigment chlorophyl, which absorbs some wavelengths of light, and have within them layers of flattened, membrane-bound sacs called thylakoids.

▶ DIATOM

A diatom is a single-celled eukaryote that uses photosynthesis to obtain food. Diatoms are classified according to their shape, and this is a pennate (long and thin) diatom. It has two long chloroplasts running along the length of its hard silica test, or shell. The chloroplasts harvest light energy and use it to turn carbon dioxide and water into simple sugars.

The **nucleus** *contains most of the cell's genetic material, deoxyribonucleic acid (DNA). DNA molecules carry the genetic information necessary for organizing the cell.*

◀ *Chloroplasts in a plant cell are found in the cytoplasm, which forms a layer close to the cell wall. The central clear region of the cell is the vacuole.*

The **rough endoplasmic reticulum** *is involved in making proteins.*

The **Golgi apparatus** *is involved in making and transporting proteins.*

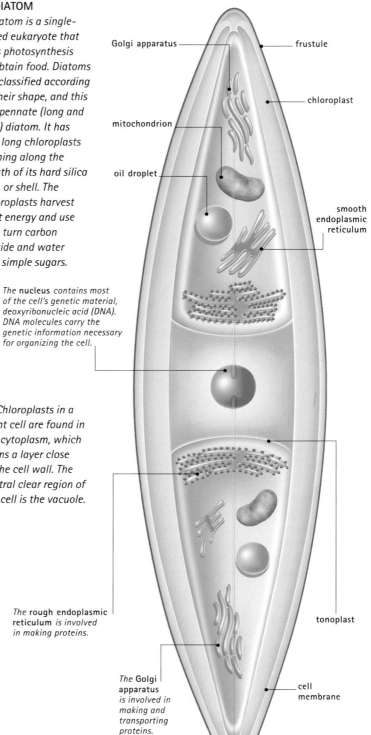

Golgi apparatus

frustule

chloroplast

mitochondrion

oil droplet

smooth endoplasmic reticulum

tonoplast

cell membrane

Movement and support

All cells have a form of internal scaffolding, called the cytoskeleton, which gives the cell its shape and serves as a highway for transporting materials from one part of the cell to another. The cytoskeleton is made of three structures: microfilaments, made of the protein actin; microtubules, formed from the protein tubulin; and intermediate filaments, made from several different proteins. These cytoskeletal elements can generate movement or transport other molecules by disassembling and reassembling in different parts of the cell. Motor molecules, called kinesin and dynein, help drive these dynamic processes.

Organisms made of many cells use muscles to move. It is the cytoskeleton of muscle cells that generates movement because the cytoskeleton contains tiny protein motors called myosin that sit on the actin microfilaments. When the signal to move arrives, the myosin ratchets along the actin, shortening the cell and making the whole muscle contract. The signal for movement coming from the nerves triggers the myosin to begin its "walk" along the actin filaments.

Single cells have developed strategies for moving in specific directions, such as away from light or toward food. Eukaryotic cells

subtubule A
bridge
subtubule B
central microtubule
outer dynein arm
radial spoke
spoke head
doublet microtubule
central sheath

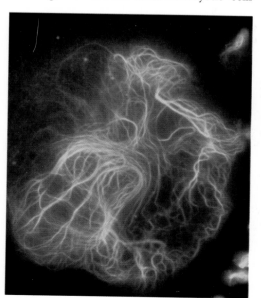

◀ These strands show some of the elements of a cell's cytoskeleton. The cytoskeleton provides both structural support to the cell and pathways along which molecular processes can occur.

▲ FLAGELLUM
Flagella and cilia are made of nine outer doublet microtubules surrounding two inner microtubules. In this illustration two of the nine doublet tubules have been removed to reveal the inner structure.

have two kinds of appendages that help them maneuver: flagella and cilia. A flagellum is a long tail that moves like a whip. Sperm cells use flagella to propel themselves as they move to find the egg. Cilia are tiny hairlike extensions that beat like tiny oars. Single-celled organisms such as the ciliate *Paramecium* use cilia to drive them through water. The cells that line the lungs of air-breathing vertebrates have cilia that help move mucus and other contaminants out of the lungs.

Both cilia and flagella have a core made of microtubules. The microtubules are arranged in a 9 + 2 array, in which nine pairs, or doublets, of microtubules are arrayed around a pair of single microtubules.

Genes and inheritance

The DNA (deoxyribonucleic acid) recipes that code for proteins are called genes. A gene is a sequence of DNA that defines what protein is to be made and when it should be made. Because proteins do many of the jobs in cells, DNA is essential for cells and organisms to survive. DNA is also the means by which this essential information is passed on to the next generation through a process called inheritance.

How is DNA information inherited? DNA forms structures called chromosomes, which generally occur in pairs. In sexually reproducing organisms, the two halves of each pair come from different parents. Simple cells such as bacteria contain one or a few chromosomes, and some plants contain more than 100 chromosomes. Humans cells contain 46 chromosomes of 23 types, organized into pairs. Cells with two copies of each chromosome are called diploid; those with one copy, such as gametes, are haploid. Cells with many copies of each chromosome are polyploid. Polyploid cells commonly occur in plants but are much rarer in animals and have been found only in one type of mammal, a species of rat that lives in Argentina.

Very often, a diploid cell will have two different copies of a single gene, one on each chromosome. The different versions are called alleles, and the cell must "decide" whether to

▼ These are human chromosomes. Humans posess 46 chromosomes, which are made of densely packed strands of DNA. The X- and Y-shaped chromosomes are those that determine sex.

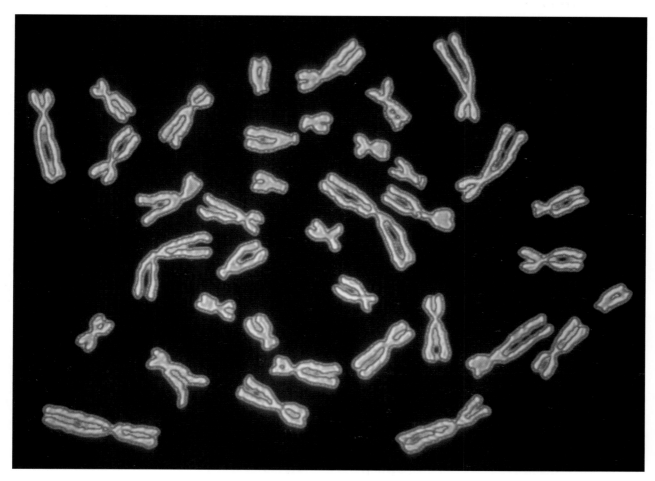

use one or the other or both. How the alleles are used and what the proteins they code for do will help determine the characteristics of a cell or organism. Scientists describe these characteristics as the organism's phenotype. The complete set of genes (including the alleles that are used and those that are not) is the organism's genotype. If the organism inherits the same allele from both parents, its genotype is homozygous. If an organism has two different alleles for a particular gene, it is heterozygous.

An Austrian monk named Gregor Mendel (1822–1884) famously conducted some

▼ ALLELES
In the mid-19th century, Gregor Mendel studied seven different traits of pea plants. He discovered that dominant and recessive alleles (different versions of the same gene) control the appearance of the seeds, pods, and stems.

GENETICS

The Human Genome Project

The Human Genome Project is an international collaboration to identify all human genes, determine the DNA sequence of the human genome, create tools for analyzing these findings, and address the ethical, legal, and social implications that may arise from the project. The United States Department of Energy and the National Institutes of Health were involved from the beginning. The Celera Corporation, a private company, joined the effort by developing a speedy way to sequence large amounts of DNA quickly and accurately using a "whole genome shotgun" technique. Human chromosomes are broken into small pieces and inserted into circular DNA strands called plasmids. The plasmids are put into bacteria, which duplicate the DNA to make sufficient quantities for sequencing. Initial descriptions of the human genome, including the surprising finding that humans may have as few as 30,000 genes, were published in 2001. The first complete sequence of the human genome was announced in 2003.

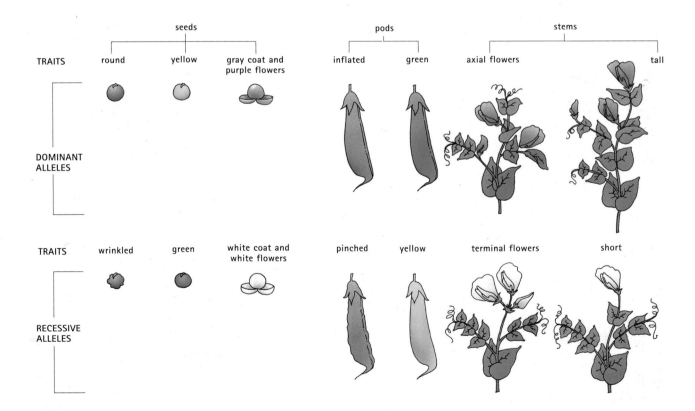

CLOSE-UP

The determination of sex

Humans are diploid animals with paired chromosomes. Thus humans have two copies, or alleles, of each gene. The exceptions are genes on the sex chromosomes. The X and Y chromosomes, so named because they are shaped like an X and a Y, help determine the sex of an individual. Individuals that inherit two X chromosomes will develop into females; those that inherit one X and one Y chromosome will become male. The Y chromosome carries a gene called SRY ("sex-determining region") Y gene. Its protein triggers a cascade of events important in male development, including the synthesis of the hormone testosterone. Because females have no Y chromosome, the SRY Y gene is never expressed and male traits do not develop.

experiments that showed that genes passed from parents to offspring help determine the offspring's characteristics. The fact that an allele is inherited in an organism's genotype does not necessarily mean it will be expressed in the phenotype. Scientists called unexpressed alleles "recessive," whereas those expressed in the phenotype were "dominant," because when present they always seemed to overrule the other allele. However, scientists are only beginning to understand the molecular mechanisms underlying dominance. Geneticists now prefer to describe an allele according to how effectively the protein it codes for perform its job. Proteins that are subnormal are designated LF (loss of function); those that work better than normal are GF (gain of function). LF alleles code for versions of proteins that for some reason do not work as well as the normal version. GF alleles code for proteins that work better than others. GF alleles either act more quickly or efficiently or do entirely new jobs that benefit the organism.

Genes that code for normal proteins are called "wild-type." When a gene fails to code for normal protein, it is called mutant. Mutant genes come in many forms. Sometimes they code for a protein that does not work, and sometimes their code fails to produce a protein

▼ RECESSIVE ALLELES

Alleles are alternative forms of the same gene. A pair of chromosomes contains two sets of alleles: one is in the chromosome derived from the father, and the other is in the chromosome derived from the mother. In this example, because both copies of one allele are faulty, the offspring will suffer from an illness, such as Tay-Sachs disease or sickle-cell anemia. In these examples, the allele is recessive, so the child will not become seriously ill if he or she inherits it from only one parent. However, the inheritance of the dominant Huntington's disease allele from just one parent is enough to pass on the illness in a full-blown form.

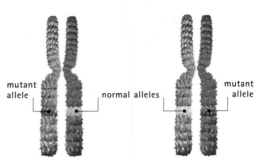

mutant allele   normal alleles   mutant allele

▲ Father's chromosome   ▲ Mother's chromosome

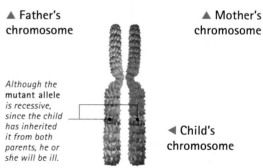

Although the **mutant allele** *is recessive, since the child has inherited it from both parents, he or she will be ill.*

◄ Child's chromosome

at all. Very occasionally, a mutant gene codes for a protein that functions, but functions differently from the wild-type.

Wild-type phenotypes are the result of millions of years of fine-tuning through evolution. Thus they usually represent a very effective solution to the task they perform: most life-forms are well adapted to their environment. The phenotypes coded for by mutant alleles are therefore almost always a disadvantage. However, when a mutant is present in a genotype, it will not necessarily be expressed in the phenotype.

For example, let us consider dwarfism, a genetic condition that causes short stature in humans and other animals. Dwarfism is caused by LF mutant alleles for growth hormone proteins, and a person must have two LF alleles to have short stature. In a person with only one LF allele and one wild-type allele, the wild-type allele produces proteins effective enough to ensure that the person is of average stature. So in dwarfism, single LF mutants are not expressed in the phenotype.

Sometimes, however, it takes the work of two normal wild-type alleles to do a job, and loss of function in either allele has a negative effect on the organism. The blood disease sickle-cell anemia is an example of this. People with one LF allele are ill but not as ill as people with two LF alleles. The LF mutants that cause sickle-cell anemia code for a hemoglobin protein that cannot bind oxygen as well as normal hemoglobin. People with one faulty allele and one wild-type can make some normal protein, but not as much as they need. Geneticists call these people haploinsufficient. A person with two LF alleles, however, will be much worse off, because none of his or her hemoglobin binds oxygen very well. GF mutant alleles produce proteins that are abnormally active. Cystic fibrosis is an example of a GF disorder, in which a GF allele that codes for a chloride channel protein allows extra chloride ions to enter a person's lung cells. Extra water also enters the cells, and they eventually burst. The ruptured cells explain why many people with cystic fibrosis have problems breathing.

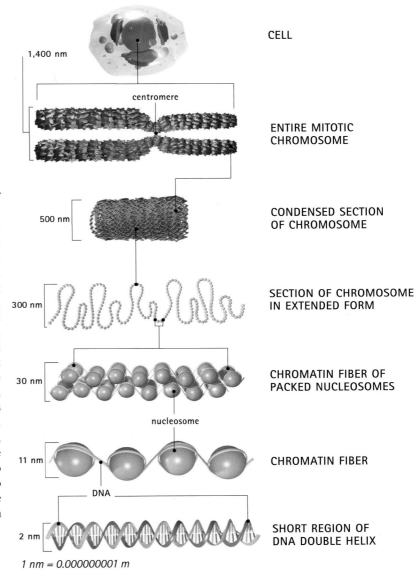

CELL

1,400 nm

centromere

ENTIRE MITOTIC CHROMOSOME

500 nm

CONDENSED SECTION OF CHROMOSOME

300 nm

SECTION OF CHROMOSOME IN EXTENDED FORM

30 nm

CHROMATIN FIBER OF PACKED NUCLEOSOMES

nucleosome

11 nm

CHROMATIN FIBER

DNA

2 nm

SHORT REGION OF DNA DOUBLE HELIX

1 nm = 0.000000001 m

▲ To pack all the genetic information contained within DNA into a single chromosome, the DNA is wrapped and folded in a number of ways. This diagram shows the method of packing at increasing levels of magnification.

IN FOCUS

The organization of DNA

There are many terms that describe how DNA is organized. Consider again the recipe analogy for genes. DNA is the individual words that make up a recipe. The recipe itself is a gene, and a group of genes are organized into a structure called a chromosome, just as a group of recipes are organized into a cookbook. DNA is tidily packaged around proteins that keep it organized, just as recipes might be arranged into chapters. This complex of DNA and protein is called chromatin. The genome is the complete collection of the chromosomes in an organism, as a library is a collection of books.

Nucleic acids: Recipes for cells

How do cells specialize? The instructions are contained in complex chemicals called nucleic acids: ribonucleic acid (RNA) and deoxyribonucleic acid (DNA). DNA makes up the "recipe book" for all proteins made by an organism. Since different organisms have different characteristics and use different strategies to survive and reproduce, each organism will have its own set of recipes. In eukaryotes, DNA is stored in an organelle called the nucleus, which is wrapped in a double layer of membrane. This nuclear membrane protects the DNA from damage that might cause errors in the recipes. Storing the DNA in the nucleus also allows the cell to control which recipes are activated—and when. When a chef cooks, he or she does not need every recipe all the time; nor does a cell. Messenger RNA (mRNA) exports a single recipe's worth of information from the nucleus as and when the protein the recipe codes for is required.

▼ *This electron micrograph shows the transcription of DNA in the bacterium Escherichia coli. Bacteria have a simple cell structure with no organelles and no nucleus, so transcription occurs in the cytoplasm (orange in the image). During transcription, mRNA strands (blue vertical lines) are synthesized when the enzyme RNA polymerase moves along the DNA strand (blue horizontal line).*

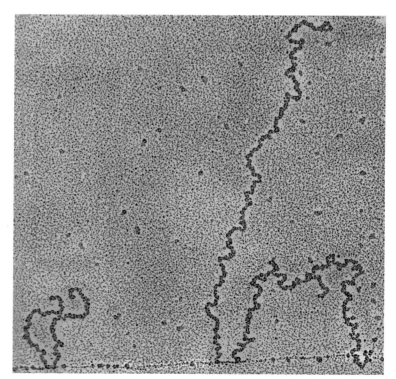

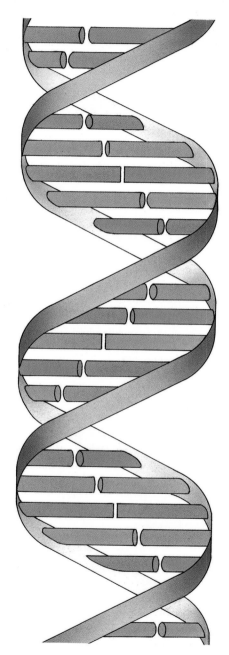

▲ **DNA DOUBLE HELIX**
The DNA double helix looks like a twisted ladder. The sides of the ladder are constructed from simple five-carbon sugars attached to a phosphate group. These "rungs" are pairs of bases, called A, C, G, and T for short. A always pairs with T, and C with G.

Think about making chocolate chip cookies. The recipe is found in a cookbook, but you want to make the cookies at a friend's house, so you copy the recipe onto a card that you can easily take with you. To make the cookies, you need the card with the recipe and all the ingredients, like flour, sugar, and chocolate chips. Then you read the recipe, add ingredients in the right order, and cook the dough to make the final product. Cells do the same thing: DNA is like the words of the cookbook containing all the recipes to make a human; mRNA is a recipe on an index card; ribosomes are the chefs; amino acids are the ingredients; and the final product is a batch of protein.

DNA and RNA are long molecules made of small units called nucleotides. Each nucleotide contains a sugar and a proteinlike molecule called a peptide, which link together to form the nucleic acid chain. In RNA the sugar is called ribose; in DNA it is deoxyribose. Each nucleotide also includes a chemical called a base. DNA molecules contain four different bases called adenine, thymine, cytosine, and guanine (or A, T, C, and G for short). DNA is more complex than RNA because each molecule consists of two chains, twisted around each other and linked by their bases. The chains pair up in a particular way. C always binds with G, and A binds with T.

The order of the bases (A's, T's, C's, and G's) in a molecule of DNA makes up the recipe that codes for a specific protein. Because of the predictable way bases pair up, a sequence of A,

CLOSE-UP

Plasmids

Most organisms store their DNA in large linear or circular structures called chromosomes. Plasmids are small circlets of DNA not included in the chromosomes. Some plasmids carry codes for making proteins that break down antibiotics; this breakdown leads to antibiotic resistance in bacteria. Bacteria can even swap plasmids by assembling a bridge, called a pilus, between them, and unwinding the plasmid so it can be threaded through the pilus to another bacterium. By this means, antibiotic resistance can be transferred to other bacteria, making a particular antibiotic less effective over time.

▼ *Plasmids are loops of DNA separate from chromosomes.*

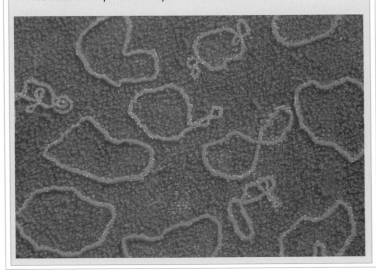

T, C, and G on one strand of DNA is complemented by a sequence of T, A, G, and C on the other strand. The recipe for a protein can be carried in one strand or the other. RNA also contains the bases A, C, and G, but it contains uracil, or U, instead of thymine, so that A binds with U instead of T.

Transcription and translation

When a cell receives a signal that a protein is needed, DNA and RNA spring into action. This signal is sent to the nucleus, and two DNA strands unravel in the spot where the appropriate gene is found. A protein called RNA polymerase then attaches to the DNA and copies the recipe by creating a strand of mRNA that matches the DNA sequence. That process is called transcription. The mRNA

IN FOCUS

The triplet code

The mRNA code that represents different amino acids is always made up of three bases, for example, AUG, CAG, and UUA. Single bases or pairs would be insufficient to code for the 20 kinds of amino acid involved in protein manufacture. Each three-base code is called a codon. There are 64 possible different three-way combinations of bases. That number is more than enough to cover the range of amino acid ingredients needed to make any protein.

strand is much shorter than the DNA strand because the DNA contains many recipes, and the mRNA contains only one.

The mRNA recipe then moves out of the nucleus and associates with a ribosome. Each group of three nucleotides in the transcribed code represents a single building block of protein, called an amino acid. Thus 300 RNA bases would code for a sequence of 100 amino acids in a protein. The ribosome reads the mRNA recipe and assembles the amino acids in the right order to create the protein. This process of converting mRNA code to sequences of amino acids is called translation.

The ribosome is the "chef," and every chef needs tools in order to cook. The ribosome uses ribosomal RNA (rRNA) and transfer RNA (tRNA) as tools in assembling proteins. The ribosome itself contains proteins and rRNA that allows it to bind to other RNA, such as mRNA and tRNA. The tRNA serves as a dictionary for translating what the mRNA

▶ DNA REPLICATION
A molecule of DNA is made up of two long strands connected by the four bases A, C, G, and T. If a base pair separates, only another partner base can fill the gap. The parent strand "unzips," and each unzipped strand re-creates its missing half. In that way, a new DNA molecule is formed. Each new molecule consists of a parent template strand and a newly synthesized ("offspring") strand. The DNA in a cell doubles in this way before the cell divides in two.

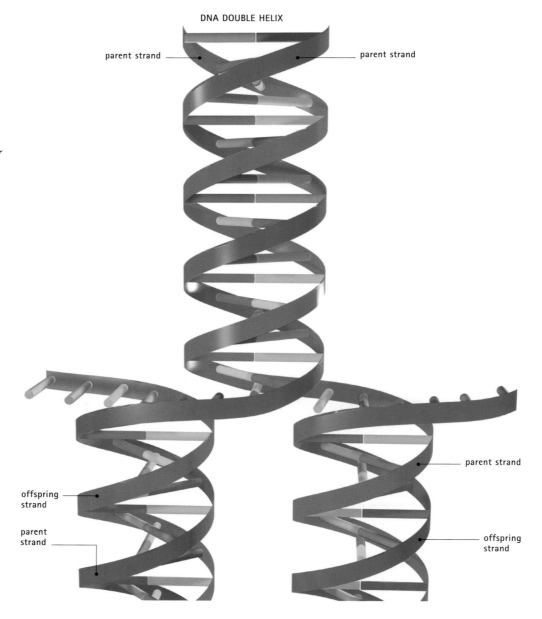

DNA DOUBLE HELIX

parent strand — parent strand

offspring strand

parent strand

parent strand

offspring strand

means in protein "language." For example, the mRNA code AUG is the instruction for including a methionine amino acid in a protein. When the ribosome reads AUG in the mRNA, it brings in a tRNA bearing the complementary base code (UAC). tRNA molecules with the code UAC are always bound to a methionine. The ribosome then strings the methionine into the protein and reads the next bit of mRNA.

Although many recipes in the DNA code for proteins by coding for mRNA, there are also recipes for rRNA and tRNA. When these recipes are read and rRNAs or tRNAs are made, they are shuttled to the appropriate spots to do their jobs, rather than aimed at the ribosomes to make proteins. rRNAs are sent to an organelle in the nucleus, called the nucleolus, for assembly into a ribosome. tRNAs are moved to the cytoplasm, where they pick up the amino acid that matches their codon and await a summons from a ribosome that needs the amino acid to make a protein.

Mitochondrial DNA

Cytoplasmic organelles such as mitochondria and plant chloroplasts have their own DNA. In almost all species, mitochondrial DNA (mDNA) is inherited only from the mother because mitochondria are carried to the next generation in eggs rather than sperm. You inherited all your mDNA from your mother, and she inherited all hers from her mother,

DNA vaccines

Vaccines are developed by introducing a foreign bit of material, called an antigen, into an organism. Antigens are typically proteins, viruses, cells, or cell parts of organisms that cause diseases, like polio or chicken pox. When immune cells find the antigen, they start an immune response. Antigens can sometimes cause illness themselves, so scientists are constantly searching for other types of vaccines. DNA vaccines are a new type of vaccine that shoots DNA from disease-causing organisms through a person's skin using air pressure. The DNA lodges in cells and codes for foreign proteins. The person's cells can then recognize foreign protein and trigger an immune response. Using DNA vaccines avoids using bacteria and viruses themselves, so there is no chance of causing disease.

and so on going back to the beginning of eukaryotic life. Lineages of mDNA never mingle. Any changes that take place happen only by random mutation. Because these mutations take place at a more or less steady rate, the number of differences between mDNA in two organisms is roughly equivalent to the amount of time that has passed since their ancestors diverged. Your mDNA will almost certainly be identical to that of your close relatives on your mother's side. However, it may differ very slightly from that of distant cousins. Scientists have used this "mDNA clock" to estimate the point at which lineages of plant and animal species diverged.

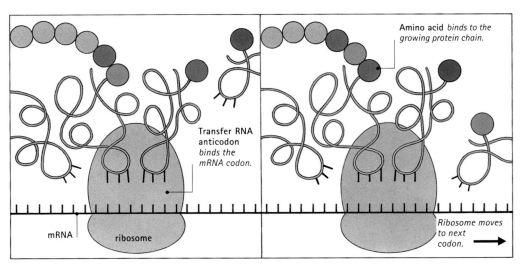

Transfer RNA anticodon binds the mRNA codon.

Amino acid *binds to the growing protein chain.*

mRNA   ribosome

Ribosome moves to next codon.

◀ **MESSENGER RNA**
Messenger RNA acts as the template that is used by ribosomes for building proteins. Each codon is made of three nucleotides and represents a particular amino acid. The codon is matched by an anticodon on the transfer RNA.

The cell cycle and cell division

▼ *Meiotic cell division is occurring in these lily pollen cells. Some of the cells have two darker blobs. These are the nuclei of the dividing cells.*

Cell division is a vital part of the growth and maintenance of a living organism, and it is essential for every new cell to receive a complete copy of the instructions inherited from the organism's parents. To ensure that each cell gets this information, a dividing cell must not only replicate itself but also duplicate its genome, and share it equally between its two offspring. The process of replication and division is called the cell cycle.

Mitosis

Before a cell divides, it is at rest and is said to be in the G_1 (G = gap) phase of interphase. When a cell prepares to divide, the chromosomes are replicated first. This occurs during a portion of interphase called the S (synthesis) phase. The cell then contains two identical copies of each pair of chromosomes. The cell enters the G_2 phase of interphase and prepares for mitosis: the M phase. During the M phase, the nuclear membrane dissolves and the chromosomes align with their duplicates in the center of the cell. The identical chromosomes are then pulled in opposite directions and the cell divides into two offspring cells. Each offspring

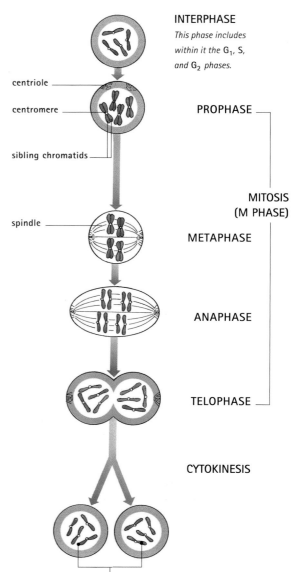

Two genetically identical diploid cells are created.

▲ THE CELL CYCLE
DNA is duplicated by chromosome replication during interphase. In prophase, the nuclear membrane breaks down, and the spindle appears. The replicated chromosomes are separated during metaphase, anaphase, and telophase to produce two identical new cells at cytokinesis.

IN FOCUS

Mitosis or meiosis?

Mitosis is the routine form of cell division in which cells replicate their DNA and divide to form two identical diploid offspring cells. Mitotic divisions take place as part of growth and routine maintenance—replacing worn-out or damaged cells. Meiosis, on the other hand, is the specialized form of division that gives rise to gametes (sex cells). In meiosis, DNA replicates once, but cells divide twice, leading to four haploid offspring cells.

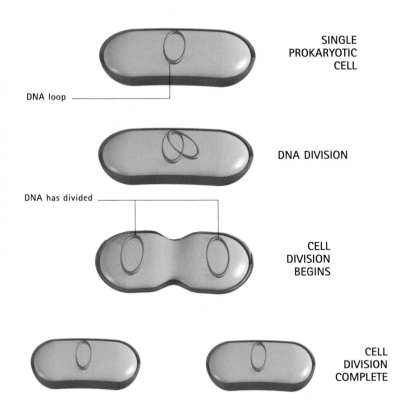

SINGLE PROKARYOTIC CELL

DNA loop

DNA DIVISION

DNA has divided

CELL DIVISION BEGINS

CELL DIVISION COMPLETE

▲ MITOSIS
Mitosis is a much simpler process in a prokaryotic cell such as a bacterium than it is in a eukaryotic cell. The loop of DNA that forms the bacteria's chromosome is duplicated, the loops separate, and the cell divides.

cell enters G_1 phase, after which it may rest or enter S phase again.

The cell cycle in prokaryotes, called binary fission (splitting in two), is much simpler. First, the organism's single chromosome is replicated and each copy is anchored on the cell membrane. The cell then begins to make new cell membrane between the two chromosomes, causing them to separate. The cell continues to make cytoplasm, cell membrane, and cell wall so that it grows to almost twice its normal size. Then it pinches down the middle and divides into two more or less equal offspring cells, with identical chromosomes.

In eukaryotes, the cell cycle is regulated by a complex series of chemical reactions. The main players in these reactions are called cyclins and Cdks, or "cyclin-dependent kinases." Kinases are enzymes that add phosphate (PO_4^{3-}) groups to proteins to activate or inhibit them. Cdks

CLOSE-UP

Inherited cancer?

Most cancers result from mutations in the DNA of somatic or body cells, such as cells in the breast or colon. Sometimes, however, mutations take place in germ-line cells (eggs or sperm). Then, they are passed on to the next generation. A germ-line mutation rarely causes cancer itself, but it can make the individual more likely to develop cancer.

and cyclins are both proteins, so each must be manufactured as it is needed.

Cyclins regulate the actions of Cdks. Therefore, Cdks cannot act unless they are bound to their partner cyclin. When the two are combined, they are called Cdk-C complexes. Each phase has its own Cdk-C complexes. G_1 phase Cdk-C complexes are made first, and trigger the cell to make proteins important for replicating DNA and for carrying out the steps of S phase. S phase Cdk-C complexes are prevented from acting until the G_1 phase is complete. S phase Cdk-C complexes initiate DNA replication. Both these steps (ending of G_1 phase by G_1 Cdk-C complexes and starting of S phase by its Cdk-C complexes) must take place before DNA replication begins. M phase, or mitotic, Cdk-C complexes are made during S phase and, again, are prevented from acting until the DNA is replicated.

Once S phase is complete, mitotic Cdk-C complexes initiate the breakdown of the nucleus and the alignment and separation of chromosomes. Mitotic Cdk-C complexes also trigger expression of another complex that ends mitosis and starts the formation of two separate cells. This final complex also helps degrade the mitotic complexes, another example of two sets of signals being required— one to stop the previous phase and another to start the next phase. All these stop and start signals help prevent the cell from continuing in the cycle prematurely.

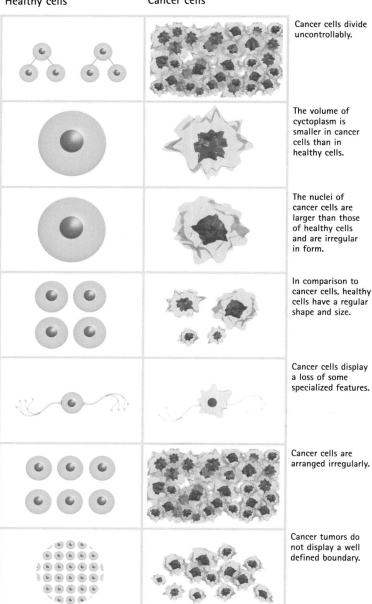

Healthy cells

Cancer cells

Cancer cells divide uncontrollably.

The volume of cyctoplasm is smaller in cancer cells than in healthy cells.

The nuclei of cancer cells are larger than those of healthy cells and are irregular in form.

In comparison to cancer cells, healthy cells have a regular shape and size.

Cancer cells display a loss of some specialized features.

Cancer cells are arranged irregularly.

Cancer tumors do not display a well defined boundary.

IN FOCUS

Natural selection

Although some mutations cause cancer, others can also be beneficial if they help the organism accomplish its two major goals: survival and reproduction. If a mutation allows an organism to survive longer or breed more successfully than organisms without the mutation it will become more common in the population as descendants of the original mutant thrive and outcompete nonmutants. This is called natural selection or "survival of the fittest." The organisms whose genes fit them best for the circumstances are "selected" for survival. The genetic diversity that allows nature to favor some lineages over others does not depend wholly on mutations. It also occurs through sexual reproduction, when chromosomes are recombined during meiosis and then blended with others to make new combinations of genes and traits.

In addition, cells have several checkpoints to ensure that each phase is complete before the cell continues to the next phase. For example, if DNA is damaged during G_1 or G_2, the cell prevents itself from entering the next phase. If the DNA is not completely replicated, the cell will stop in S phase. Finally, if the chromosomes do not line up properly before separating, the cell will halt in M phase. The checkpoints are controlled by proteins coded by tumor suppressor genes. They are called tumor suppressor genes because if they are damaged, the malfunctioning cells divide uncontrollably and form tumors. In fact, all these checkpoints and start and stop signals help prevent errors in cell division and thus help prevent the growth of cancers.

Occasionally, one of these checkpoints or Cdk-C complexes fails because the genes

◀ RECOGNIZING CANCER CELLS
Doctors are able to identify cancer cells because they display a number of significant differences in form and arrangement compared to healthy cells. The differences include an enlarged nucleus, an irregular arrangement of cells, and an unusual variation in cell size and shape.

involved have mutations. The cell divides uncontrollably and passes this ability on to its offspring cells. Usually several mutations must occur before cancer forms; and because mutations accumulate as an individual ages, older people are more likely to develop cancer than the young. Loss-of-function mutations in tumor suppressor genes are one of the leading causes of cancer because they can make tumor suppressor genes neglect their checkpoint duties. Other genes stimulate cells to divide. If these genes develop gain-of-function mutations, they too can cause cancer by stimulating cells to divide uncontrollably.

Most cells continue to undergo mitotic division on a regular basis as part of routine maintenance of the living tissues. However, some—notably most vertebrate neurons, skeletal, and cardiac muscle cells—have exited the cell cycle permanently and can no longer divide. Some continue to grow as an organism grows, but they do not increase in number. Since these cells no longer replicate themselves, they cannot be replaced if they are severely damaged or destroyed, as by a stroke or spinal cord injury. Researchers are trying to develop ways of stimulating neurons and muscle cells to regenerate.

Meiosis

Eggs and sperm (female and male sex cells, or gametes) are haploid. These cells carry half the DNA usually needed to make an offspring. Haploid gametes form by a special kind of cell division called meiosis. To create an egg cell, for example, the chromosomes within a cell are replicated. Normal diploid human cells have 46 chromosomes, 23 from each parent. Once the chromosomes are replicated, the cell has 92 chromosomes in total. Then the chromosomes line up on opposite sides of the cell and swap fragments. The DNA on both sides becomes mixed, or recombined. The chromosomes then separate and the cell divides, creating two diploid cells, each with 46 chromosomes. These cells divide again to create four haploid gamete cells, each with 23 unique chromosomes. This is how the genetic information your parents inherited from your grandparents was remixed and shared out to create the egg and the sperm that became you. The recombination of chromosomes is what gave you the particular

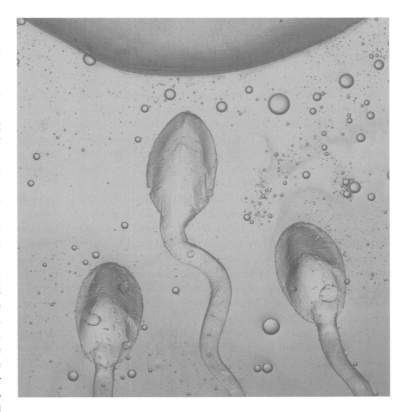

set of 46 blended genes that make you unique (unless you are an identical twin).

Apoptosis

Some cells commit "suicide" through a process called apoptosis. Usually this happens when the organisms is stressed by lack of nutrients or when the cells are damaged. Other cells, like those in the immune system, determine that certain cells should be "murdered" because they are foreign or have been infected by a virus. Both of these processes are different from cell death caused by an injury. Apoptosis is series of well-defined steps that prevent damage to neighboring cells. Cell death caused by injury is uncontrolled and also leads to damage of neighboring cells.

ERIN DOLAN

▲ *Three human sperm approach a human egg. Sperm and eggs are haploid gametes. If one of the sperm penetrates the egg cell and fertilizes the egg, a diploid cell will result. This will contain the genetic information from both parents.*

FURTHER READING AND RESEARCH
Marshall Graves, Jenny. 2004. *Sex, Genes, and Chromosomes.* Cambridge University Press: Cambridge, UK.
Snedden, R. 2003. *Cell Division and Genetics.* Heinemann Library: Chicago, IL.

Circulatory system

Around 1.5 billion years ago, some single-celled organisms began to band together, first into colonies and later into multicellular organisms, or metazoans. Only the outer surface or the exposed areas of internal cavities were in direct contact with the environment. This new arrangement presented a serious problem for the transport of oxygen and nutrients to cells farther away from such surfaces. The rate of diffusion (movement of molecules into or from a structure) is severely limited by the distance a molecule must travel. One solution to the problem was the evolution of circulatory systems that carried fluid between the body surface and the tissues inside, taking oxygen and nutrients in one direction and wastes in the other. In humans and other vertebrates, this fluid takes the form of blood, which is carried in blood vessels and pumped by a muscular organ, the heart.

Not all animals deal with the challenge of getting material to and from cells in that way, however. Many small aquatic animals still rely on diffusion through their outer surface to move nutrients, gases, and wastes, as did their ancient metazoan ancestors. The outer surface of such animals must be thin, otherwise the diffusion distance becomes too great. The animals also need to maximize the surface area available for such exchanges by being flattened, multibranched, or leaf-shaped, for example.

The diffusion of dissolved oxygen from seawater into a jellyfish takes place on both sides of its swimming bell. Other animals take this further, and have a dedicated, blind-ending internal cavity, or gastrovascular cavity, through which absorption or waste disposal can take place. The gastrovascular cavity may be highly branched to increase the available surface area still further. For example, hydras (cnidarians, relatives of the jellyfish) have a gastrovascular cavity that extends throughout the body and into the tentacles. The cavity serves the dual function of absorption of food and distribution of substances around the body. No hydra cell is more than one cell away from either the cavity or the water outside the animal.

▼ Fruit bat

Fruit bats have a mammalian circulatory system with a four-chamber heart and a network of arteries and smaller vessels called arterioles and capillaries, which together carry oxygen-rich blood to all parts of the body. Oxygen-depleted blood travels from body cells in veins.

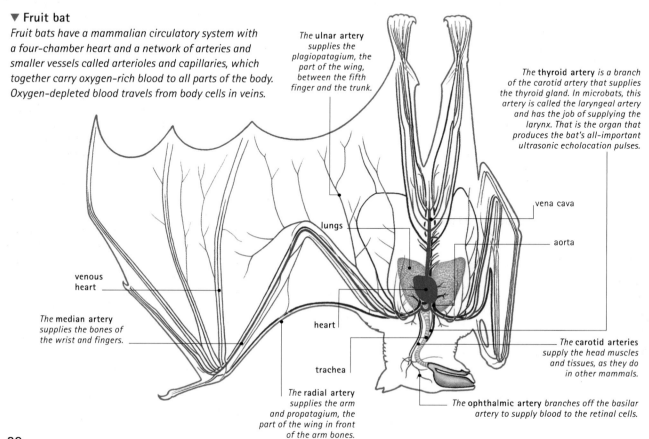

The **ulnar artery** supplies the plagiopatagium, the part of the wing, between the fifth finger and the trunk.

The **thyroid artery** *is a branch of the carotid artery that supplies the thyroid gland. In microbats, this artery is called the laryngeal artery and has the job of supplying the larynx. That is the organ that produces the bat's all-important ultrasonic echolocation pulses.*

vena cava

lungs

aorta

venous heart

The **median artery** *supplies the bones of the wrist and fingers.*

heart

The **carotid arteries** *supply the head muscles and tissues, as they do in other mammals.*

trachea

The **radial artery** *supplies the arm and propatagium, the part of the wing in front of the arm bones.*

The **ophthalmic artery** *branches off the basilar artery to supply blood to the retinal cells.*

Open circulatory systems

For efficient diffusion, a hydra may be little more than two cells thick. However, simple diffusion cannot supply the demands of larger animals made up of thicker layers of cells. Cells of such animals are surrounded by a fluid that brings nutrients and oxygen to the cells and takes away wastes. Circulatory systems evolved to move this fluid around.

Almost all larger animals have a muscular pump that drives blood around the body. Insects have a long structure called a dorsal vessel running close to the top (dorsal) side of their body. The hind part of the vessel, the heart, contracts, forcing fluid (called hemolymph) forward into a noncontractile tube, the aorta. This leads to smaller vessels that empty around the tissues. The fluid may be given a boost by extra pumping organs at the base of the legs and antennae (feelers).

After supplying nutrients, the hemolymph passes into the body cavity, or hemocoel. Assisted by muscular movements of the abdomen, the hemolymph moves through the celom before being drawn back into the dorsal vessel through valves called ostia. Systems like this, where fluid flows freely through the body cavities, are called open-circulation systems.

Insects do not rely on their hemolymph to deliver oxygen to cells. They instead have a system of air tubes called a trachea. The efficiency of this system allows insects to have a high metabolic rate (rate of energy usage), so they can perform energetically demanding activities such as flight. However, for other open-circulation animals, such as bivalves, gastropods, and most other mollusks, life is much slower. Their blood (like that of insects) moves slowly, but these animals depend on it to pick up oxygen at a respiratory surface (a gill or lung) before transporting it around the body. An open circulation does not deliver blood (and hence oxygen and nutrients) quickly enough to the cells to allow high metabolic rates.

SYSTEM HIGHLIGHTS

THE BLOOD This fluid is responsible for carrying nutrients, gases, and wastes around the body. *See pages 32–33.*

THE VERTEBRATE HEART Fish have a two-chamber heart; mammals, birds, and most reptiles have a four-chamber heart. *See pages 34–36.*

HOW HEARTS FUNCTION The heart is the driving force of the circulatory system. It pumps blood around the body while keeping perfect time. *See pages 37–41.*

ARTERIES, VEINS, AND CAPILLARIES The vessels through which blood travels around the vertebrate body. *See pages 42–44.*

REGULATION AND CONTROL The body carefully controls the circulatory system. *See page 45.*

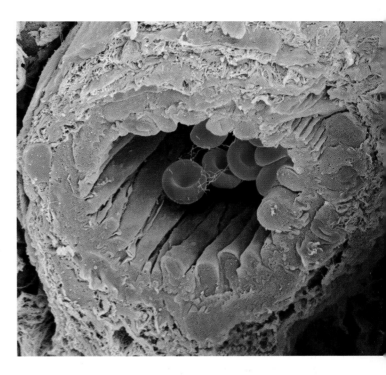

▲ *A scanning electron microscope image of a small human artery called an arteriole. Red blood cells are visible in the central area. These cells pick up oxygen from the lungs and transport it to the body's tissues. Magnified 1,430 times.*

Closed circulatory systems

Squid are mollusks, but they have a much higher metabolic rate than other mollusks such as slugs and snails. Squid are able to sustain a higher metabolic rate because they have evolved a different type of circulatory system—a closed one. In contrast to open systems, where blood and tissue fluid are one and the same, these fluids are separate in squid and other animals with a closed system. Blood is contained in vessels throughout the circulation and is driven around by one or more hearts. Some parts of the blood never leave the vessels, while tissue fluid bathes the cells. Blood can flow much faster through a closed circulation, so tissues can be supplied more quickly, allowing a higher metabolism. Also, the resistance of the vessels to blood flow can be altered, so extra blood can be delivered to specific areas of the body when needed, or the supply can be reduced. Closed circulatory systems occur in most larger animals, including all vertebrates. Many invertebrates have them, too. Earthworms have two major blood vessels, one dorsal and one ventral (running close to the animal's underside). These are connected in each segment by a network of smaller vessels. A series of five hearts near the head keep the worm's blood circulating.

The blood

CONNECTIONS

COMPARE the blood of a vertebrate such as a lion with that of an insect such as a *HONEYBEE*. Vertebrate blood is red, owing to the presence of hemoglobin; insect blood does not carry oxygen and is clear.

Blood is the fluid that flows through the vertebrate circulatory system. Blood is responsible for carrying materials from one part of the body to another, delivering oxygen and nutrients to cells and taking away their wastes.

What does blood contain?

Blood is made up of a yellowish fluid called plasma in which two types of blood cells are suspended. Red blood cells carry gases, and white blood cells engulf infectious organisms, destroy rogue cells, and regulate immune responses. Plasma also contains platelets, cell fragments that help blood clot. Plasma is about 90 percent water, with nutrient molecules, dissolved gases, tiny charged particles called ions, and chemicals such as hormones making up the rest. Plasma also contains an important group of proteins called plasma proteins. These have many roles: they regulate the water balance of the plasma; help carry materials (such as iron, absorbed in the gut and vital in red blood cell production); trigger clotting; and stimulate the body's defenses against foreign objects.

Oxygen carriers

Red blood cells are astonishingly numerous. At any one time, trillions are circulating around a person's bloodstream; they make up about 45 percent of the blood and give it its color. Their sole function is to carry gases (mostly oxygen, though they help return waste carbon dioxide to the lungs) around the body.

Red blood cells possess an iron-containing protein called hemoglobin. When a cell enters blood vessels in the lungs, oxygen binds to the hemoglobin. The oxygen is released when the cell enters capillaries in the rest of the body.

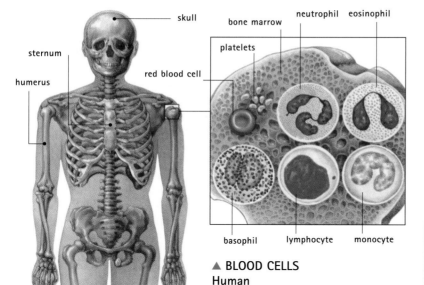

Labels: skull, bone marrow, neutrophil, eosinophil, sternum, platelets, humerus, red blood cell, femur, basophil, lymphocyte, monocyte

▲ BLOOD CELLS
Human
Blood cells are made in a variety of different places all over the body. Most are made in the skull, spine, ribs, sternum, and the ends of larger bones such as the humerus and femur. There, bone marrow produces all the cell types: red blood cells, white blood cells, and platelets.

IN FOCUS

How blood cells are made

All blood cells form in the bone marrow, especially that of long bones such as the femur. Blood cells develop from amazing cells called stem cells. These are always dividing, sometimes forming other stem cells, but they also have the ability to form any body cell, including the blood cells. Blood cell production is stimulated by a hormone, erythropoietin, which is made mainly by the kidneys. The rate of blood cell production is affected by oxygen levels in the body; if these are low, extra erythropoietin is released. This causes the stem cells to produce more red blood cells and speeds up their maturation.

Red blood cells are so dedicated to their role that in mammals, mature cells do not have nuclei (control centers), so they cannot divide like most other cells. The red blood cells use the space saved to pack in more hemoglobin. They need to be very flexible to move through narrow capillaries. Red blood cells are disk-shaped and have hollows on the top and bottom; this unusual shape increases the surface area available for gas exchange.

White blood cells

White blood cells occur not only in the blood, but also in other parts of the body, such as the lymphatic system. There are five different types. Neutrophils attack other cells, such as bacteria or rogue body cells, which they engulf and break down before splitting apart. The fluid released may collect near the site of infection and is called pus. Eosinophils specialize in attacking parasites, especially the young of parasitic worms. Basophils respond to particles that cause allergic reactions. Monocytes remove dead cell debris. The last white blood cell group, lymphocytes, drive the specific immune response, whereby infectious organisms are recognized and destroyed.

Plasma also contains small fragments of cells called platelets. Like blood cells, these are created in the bone marrow. Platelets form a plug to stop blood loss from damaged tissue. They then release chemicals that lead to coagulation, in which a protein clot seals the wound. Platelets are generally inactive, but when they contact damaged tissue their surface structure alters. This response makes them sticky, so they bind to the damaged tissue and begin the coagulation process.

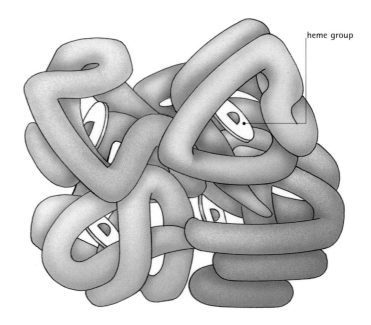

heme group

▲ HEMOGLOBIN MOLECULE

The molecule is made up of four chains, each of which has one heme group, the iron-making molecule that transports oxygen and gives blood its red color.

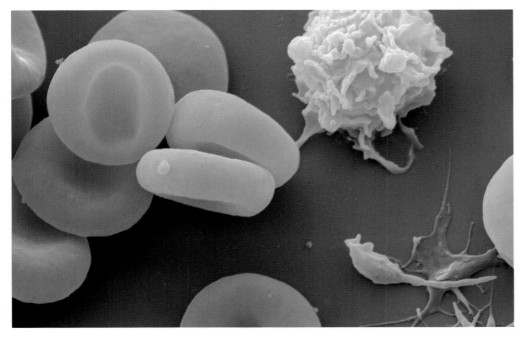

◄ In mammals, mature red blood cells do not have a nucleus. The white blood cells are part of the immune system and help protect the body against disease. The green structures are platelets, which are fragments of cells and are involved in the process of clotting.

The vertebrate heart

The number of chambers in an animal's heart varies depending on the type of animal. Fish have a two-chamber heart; amphibians such as frogs have a three-chamber heart; and mammals, birds, and most reptiles have a heart with four chambers. In reptile hearts the lower chambers, or ventricles, are usually not completely separated.

All vertebrates have closed circulatory systems, with blood moving at relatively high pressure around the body while contained in vessels. Vertebrate hearts have two or more separate sections, called chambers, connected by valves. Vertebrate blood carries nutrients to cells and takes away wastes. The blood also carries oxygen to cells and waste carbon dioxide to the respiratory surface. In fish and young amphibians, the respiratory surface takes the form of gills. In other vertebrates the respiratory surfaces are lungs.

Fish hearts

A fish heart has two chambers. One, the atrium, is relatively thin walled. It receives blood from the body via veins and pumps it on to a more muscular chamber, the ventricle. This chamber pumps the blood to the gills, where wastes move into the water, and oxygen moves in. The oxygen binds to hemoglobin, a substance carried by red blood cells. Blood leaving the gills collects in a large vessel, the aorta. From there, blood passes along a succession of narrower vessels (arteries and arterioles), until it reaches the tissues. Moving into tiny, narrow-walled vessels called capillaries, the blood discharges its cargo of oxygen and nutrients, and picks up wastes in return. The blood then moves through venules and veins before passing back to the heart.

Fish blood leaves the heart at high pressure, but much of this is lost when the blood is forced through the narrow vessels of the gills.

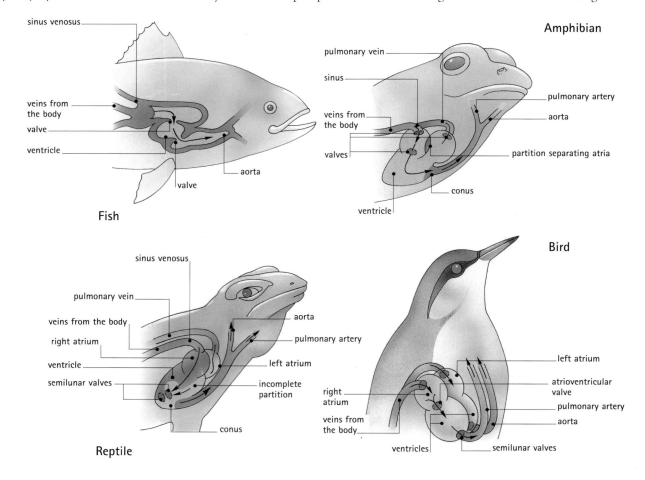

sinus venosus

veins from the body
valve
ventricle
aorta
valve

Fish

Amphibian

pulmonary vein
sinus
veins from the body
valves
pulmonary artery
aorta
partition separating atria
conus
ventricle

sinus venosus
pulmonary vein
veins from the body
right atrium
ventricle
semilunar valves
aorta
pulmonary artery
left atrium
incomplete partition
conus

Reptile

Bird

right atrium
veins from the body
ventricles
left atrium
atrioventricular valve
pulmonary artery
aorta
semilunar valves

Hagfish hearts

Lampreys and hagfish resemble the earliest vertebrates. They differ from other vertebrates in many features, such as the absence of jaws; their circulatory systems are similarly unusual, being "semi-open." They have some blood vessels, notably an aorta that runs along the midline of the body. However, as well as being pumped into vessels, blood also "leaks" from the heart, which, relative to that of most other vertebrates, is massive. Hagfish hearts lack a direct blood supply, and rely on this leakage for nutrients. However, blood that arrives in this way lacks oxygen, so contraction of hagfish heart muscle is anaerobic (does not use oxygen).

Thus, fish blood is at a low pressure for most of its journey around the body. The low pressure limits the rate at which oxygen can be supplied to tissues such as muscles.

The evolution of some fish from life in water to life on land was accompanied by important changes in their circulatory system. African lungfish are similar to those ancient ancestors. The pools in which lungfish live often become low in oxygen or dry up completely. The fish cope by having a pouch in the gut that acts as a lung. Many thin blood vessels pass through this pouch to pick up oxygen. The blood vessels are supplied by a pair of arteries that once connected to a pair of gills; a new vessel carries blood back to the heart. Another pair of gills no longer functions; their arteries now lead from the heart directly to the aorta, bypassing the gills.

The lungfish atrium is part-divided; the left side receives oxygen-rich blood from the lung, and the right side receives deoxygenated blood from the body. The two blood flows mix a little as they pass into the ventricle, but mostly oxygenated blood passes along the gill arteries to the aorta, and mostly deoxygenated blood goes to the lung and functional gills. These features help separate blood moving to the lung (the pulmonary circulation) from the flow around the body (systemic circulation).

Three chambers

Keeping pulmonary and systemic flows separate became increasingly important as vertebrates took to life on land. Because of separation, blood pressure was no longer reduced by being forced through the gills on its way to the rest of the body; instead, oxygen-rich blood could remain at high pressure all the way around the body.

Modern amphibians such as frogs and salamanders have two atria and a ventricle. The ventricle pumps blood to the lungs and to the rest of the body. One atrium receives blood from the lungs, and the other atrium receives blood from the body. Both deliver their blood to the ventricle, but mixing is limited by valves in the ventricle. They direct deoxygenated

▼ **CIRCULATORY SYSTEMS**

The five groups of vertebrates display increasing complexity in the structure of the heart and circulatory system. The increase in complexity is paralleled by an increase in efficiency.

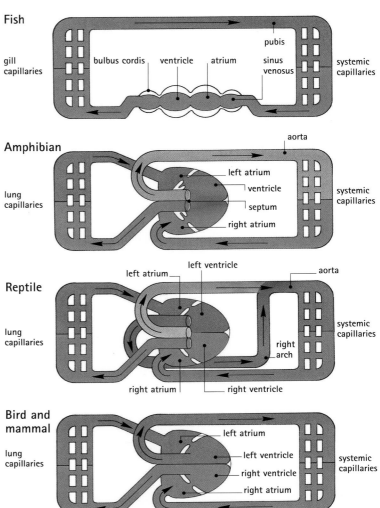

Fish

gill capillaries · bulbus cordis · ventricle · atrium · sinus venosus · pubis · systemic capillaries

Amphibian

lung capillaries · aorta · left atrium · ventricle · septum · right atrium · systemic capillaries

Reptile

lung capillaries · left atrium · left ventricle · aorta · right atrium · right ventricle · right arch · systemic capillaries

Bird and mammal

lung capillaries · left atrium · left ventricle · right ventricle · right atrium · systemic capillaries

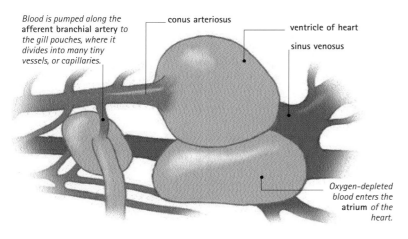

Blood is pumped along the **afferent branchial artery** *to the gill pouches, where it divides into many tiny vessels, or capillaries.*

conus arteriosus

ventricle of heart

sinus venosus

Oxygen-depleted blood enters the **atrium** *of the heart.*

▲ MAIN HEART
Atlantic hagfish
Oxygen-depleted blood passes through the four chambers of the heart: the sinus venosus, atrium, ventricle, and conus arteriosus. Blood pressure is raised as it is pumped through. The blood then passes into tiny blood vessels in the gill pouches where gas exchange occurs. Oxygen passes into red blood cells in the bloodstream, and carbon dioxide is removed from the blood.

blood to the lungs and oxygen-rich blood to the aorta, and from there to the rest of the body.

Most reptiles also have a three-chamber heart with valves to separate the circulations, but their heart is more complex than that of amphibians. Rather than a single aorta, reptiles have two aortas extending from the ventricle. Blood on the right side of the ventricle passes to the lungs, just before the left part contracts. With the vessel to the lungs already filled, blood from this part of the heart is forced

into the aortas and the systemic circulation. Reptiles are capable of fast bursts of activity, but they can also be inactive for long periods. During inactivity, breathing takes place only occasionally. When a turtle is not breathing, it can save energy by allowing its blood to bypass the pulmonary circulation altogether. It does this by constricting blood vessels in the lungs. Resistance increases in the pathway through the lungs, so blood instead passes into the aortas and from there to the rest of the body.

Birds and mammals have a four-chamber heart. However, systemic and pulmonary circulations are completely separate in these animals, which have two atria and two ventricles. Oxygenated and deoxygenated blood can never mix, and so the tissues receive as much oxygen as possible. Blood with the lowest oxygen content and highest carbon dioxide content always passes to the lungs. The two circuits can also run at different pressures. Bird and mammal tissues have high nutrient and oxygen demands, requiring lots of narrow capillaries and, therefore, a high-pressure systemic system to force blood through. The lungs have fewer capillaries, so their blood supply can be kept at a far lower pressure.

CLOSE-UP

Complex crocodiles

Unlike other reptiles, crocodiles have a four-chamber heart. However, the valves at the top of the right ventricle in Australian estuarine crocodiles are unique. Flaps of connective tissue on their surface can intermesh; they are called teeth owing to their resemblance to the teeth of a cogwheel. When the crocodile is relaxed the teeth mesh together; most of the blood pumped out by the right ventricle goes to the body. However, when the animal is under stress the hormone adrenaline causes the teeth to contract. This allows more blood to be pumped to the lungs, where it picks up oxygen. The oxygenated blood is then pumped to the tissues, enabling them to sustain more vigorous activity.

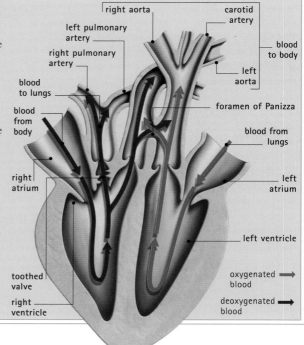

right aorta

carotid artery

left pulmonary artery

right pulmonary artery

blood to body

blood to lungs

left aorta

blood from body

foramen of Panizza

blood from lungs

right atrium

left atrium

left ventricle

toothed valve

oxygenated blood →

right ventricle

deoxygenated blood →

How hearts function

The driving force behind the circulatory system is the heart. The human heart is located just to the left of the center of the chest cavity, encased by a protective membrane, the pericardium. As in all other mammals and in birds, there are four chambers in the human heart. The atria are relatively narrow walled and serve as collection points for blood. The more muscular ventricles pump blood along arteries, either to the lungs or around the body.

Into the heart

Blood enters the heart via a system of increasingly large veins, culminating with the inferior and superior venae cavae. These veins collect blood from the lower and upper parts of the body respectively, and they empty into the right atrium of the heart. At this stage, the blood has already yielded its oxygen to the tissues of the body and is laden with high levels of waste carbon dioxide. The blood must travel to the lungs to be replenished with oxygen and to release the waste gas.

The blood flows passively from the right atrium to the right ventricle, while some of it is pumped in actively by the contraction of the chamber's walls. The right ventricle then contracts, forcing blood into the pulmonary artery. This artery immediately branches into two forks, one for each of the lungs. This part of the circulatory system is called the pulmonary circulation.

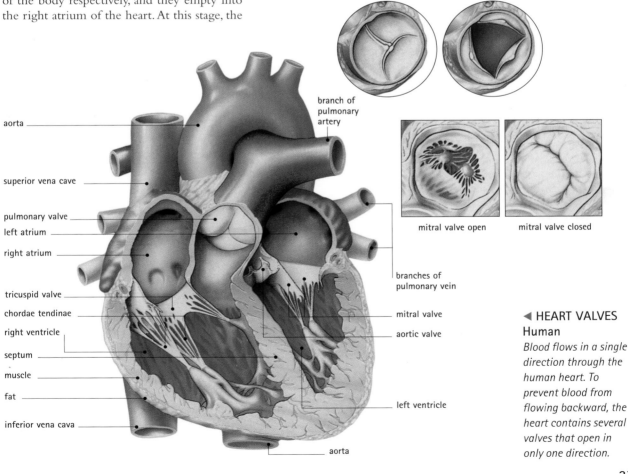

pulmonary valve closed

pulmonary valve open

branch of pulmonary artery

mitral valve open

mitral valve closed

aorta

superior vena cave

pulmonary valve

left atrium

right atrium

tricuspid valve

chordae tendinae

right ventricle

septum

muscle

fat

inferior vena cava

branches of pulmonary vein

mitral valve

aortic valve

left ventricle

aorta

◄ HEART VALVES
Human
Blood flows in a single direction through the human heart. To prevent blood from flowing backward, the heart contains several valves that open in only one direction.

Be still thy beating heart

Close your eyes and listen hard after doing some vigorous exercise and you will hear your heartbeat. What causes this familiar sound? You might think the heartbeat is caused by the contracting of the ventricles, but it is not actually caused by muscular activity at all. It is caused by the shutting of the valves that separate atria from ventricles. The sound made by the human heart is typically described as "lubb–dubb."

In the first stage of the heart cycle, the ventricles fill with blood (1) and then contract, causing the blood to press against the mitral and tricuspid valves, producing the first sound, "lubb" (2). The blood then leaves the ventricles (3) and presses back on the pulmonary and aortic valves, producing the second sound, "dubb" (4). The atria fill once more (5) and the cycle begins again.

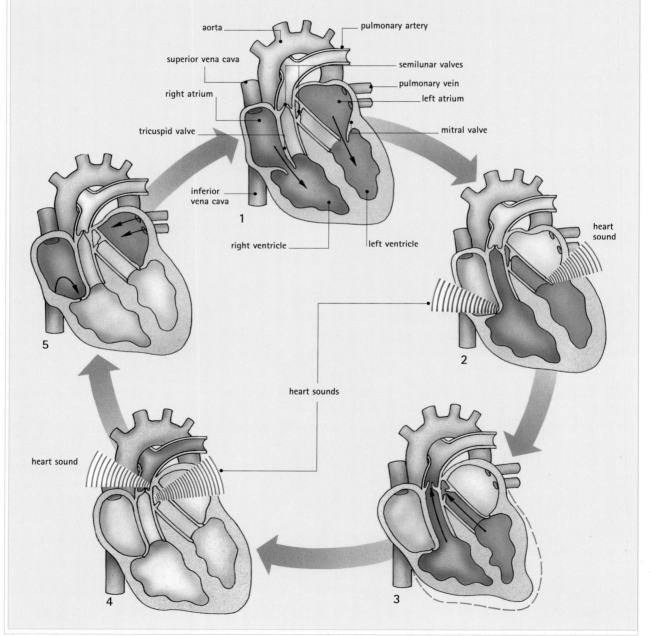

Back to the heart

Rich with oxygen after traveling to the lungs, the blood returns to the heart along the pulmonary veins. This time the blood enters the left atrium. Now the blood needs to be forced around the rest of the body. It must enter the systemic circulation, so that oxygen and nutrients can be delivered to the tissues that need them.

The blood flows from the left atrium to the left ventricle, again mainly passively. When blood moves from either of the atria to their corresponding ventricles, backflow is avoided by the atrioventricular valves. The mitral valve prevents blood from flowing back into the left atrium from the left ventricle, and the tricuspid valve prevents blood from flowing back into the right atrium from the right ventricle.

The walls of the left ventricle are formed by banks of powerful muscles. The walls squeeze until the pressure of the blood forces open another valve, the aortic valve. Blood moves through the valve into the largest blood vessel in the body, the aorta. The left ventricle is much more muscular than the right. It needs to be. Blood traveling to the lungs needs only a low pressure to allow it to complete its circuit. Blood entering the systemic circulation from the left ventricle must pass through dense beds of capillaries, which provide great resistance to the flow. The left ventricle must therefore exert a high pressure on the blood to force it right around the body and back to the heart.

Supplying the heart

The blood travels along the aorta, and is then carried by smaller arteries to the rest of the body. Two important vessels, the coronary arteries, branch off from the aorta while still inside the heart. They pass out of the heart, looping across the surface to supply the heart muscles themselves.

Cardiac (heart) muscle cells need lots of energy to keep going and so require a rich supply of nutrients and oxygen. The energy is delivered via a vast number of capillaries, thin-walled vessels through which molecules can pass to and from the blood. Unlike most other muscles, cardiac muscle cannot function for long without oxygen. Anything that interrupts the flow of oxygen to the heart can quickly lead to damage. This is what happens during a heart attack, when the cardiac arteries may become blocked.

How do contractions happen?

Cardiac muscle has some unusual properties vital to its effectiveness as a pump. Muscles generally contract when they are stimulated by an electrical pulse from the nervous system, originating in the brain or spinal cord. Cardiac muscle also contracts in response to electrical pulses, but these pulses do not stem from the nervous system. Instead, the heart initiates

COMPARATIVE ANATOMY

Efficient birds

Flying is extremely demanding in terms of energy. Birds in flight need to deliver lots of oxygen to their wing muscles. They have evolved superefficient lungs to help them do this. Their heart is similarly efficient. Birds have larger hearts relative to body mass than mammals. Hummingbird hearts are the largest relative to size, since hovering requires vast amounts of energy. Bird hearts also pump more blood than those of mammals. They tend to beat more slowly than that of mammals of equivalent size, but deliver more blood with each beat.

▼ VENOUS HEARTS
Bat
In bats, the beating of the wings creates a centrifugal force. This force pushes the blood to the wing tips. To counteract this effect, bats have rhythmically contracting veins called venous hearts that pump blood back to the main heart.

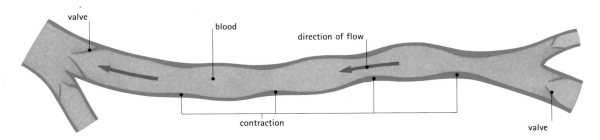

valve · blood · direction of flow · contraction · valve

▲ *The mitral valve of a human heart separates the left atrium from the left ventricle and allows only a one-way flow of blood from atrium to ventricle. The stringlike chordae tendineae attach the outer parts of the valve to projections in the wall of the ventricle.*

electrical activity and contractions itself, without nervous input. These electrical pulses (or action potentials) are able to spread throughout the whole heart, because each heart cell is in electrical contact with all of its neighbors through connections called gap junctions. An action potential started by a cluster of cells (called a pacemaker) travels swiftly, stimulating other heart cells to contract in a tightly synchronized wave. Another important property of cardiac muscle cells is the fibers that extend from each cell into its neighbors. The fibers create a tough linkage, allowing powerful contractions to take place that do not tear the cells apart.

The path of electrical activity

Now let us follow the path of electrical activity as it moves through the heart to trigger the heartbeat. The heartbeat is started by a pacemaker found at the junction between the right aorta and the superior vena cava—the point at which much of the blood first re-enters the heart. This cluster of cells is called the sinoatrial (or S-A) node. The node cells fire out an electrical impulse that passes from cell to cell, across the walls of both the left and right

atria. This impulse causes the two chambers to contract gently and simultaneously. Blood then moves from the atria to the left and right ventricles respectively.

There are no gap junctions across which action potentials can jump from the atria to the ventricles. The chambers are separated by insulating connective tissue through which the action potential cannot pass. The impulse started by the S-A node does, however, pass into another clump of pacemaker cells called the atrioventricular (or A-V) node, which lies just above the ventricles. It is this second cluster of cells on which the contractions of the ventricle depend.

The A-V node fires out a second action potential. This passes along a bundle of specialized fibers, which soon divides into two—one branch for each ventricle. The branches then separate into structures called Purkinje fibers. The fibers penetrate and run through the walls of the ventricles, allowing the action potential to trigger the contraction of the ventricular muscle cells swiftly and in unison. The ventricles squeeze, forcing blood either to the lungs or around the systemic circulation.

Passing the action potential from the atria to the ventricles by the A-V node requires delicate timing. There is a short delay between the arrival of the action potential at the A-V

CLOSE-UP

The long route back

The capillary beds of most tissues open into veins that head straight back to the right atrium. However, blood draining from the spleen, stomach, and intestines takes a detour, and heads for a second set of capillary beds in the liver. This is called the hepatic portal system. Through these capillaries, the liver absorbs surplus materials taken in across the wall of the gut. For example, glucose is removed and converted to an easily stored form, glycogen. Other molecules, such as amino acids and toxins, are taken in to be broken down. The blood then leaves the hepatic portal system and returns to the general circulation.

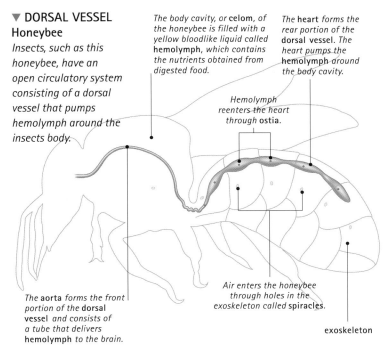

▼ DORSAL VESSEL
Honeybee
Insects, such as this honeybee, have an open circulatory system consisting of a dorsal vessel that pumps hemolymph around the insects body.

The body cavity, or celom, of the honeybee is filled with a yellow bloodlike liquid called hemolymph, which contains the nutrients obtained from digested food.

The heart forms the rear portion of the dorsal vessel. The heart pumps the hemolymph around the body cavity.

Hemolymph reenters the heart through ostia.

The aorta forms the front portion of the dorsal vessel and consists of a tube that delivers hemolymph to the brain.

Air enters the honeybee through holes in the exoskeleton called spiracles.

exoskeleton

node and the release of the second action potential to stimulate the ventricles. This delay explains why the atria contract before the ventricles. The delay allows a wave of contraction along the heart to take place, moving blood from the atria to the ventricles and out into the arteries smoothly and in a precisely regulated sequence.

IN FOCUS

Beating out of time

The ventricles normally rely on the atrioventricular node to regulate their contractions. They can beat even when this pacemaker is damaged; however, there is then a chance that ventricles will contract irregularly. They may then begin to twitch randomly. This is called ventricular fibrillation. Blood flow halts and death is swift. Hospital emergency wards are equipped with machines called defibrillators. They fire a burst of electrical current into a patient, jolting the heart back into its normal rhythm and possibly saving a person's life.

▼ THROMBOSIS
Human
At left, the blood flow is partly blocked by fatty deposits. At right, a blood clot, or thrombosis, has caused a total blockage in the already narrowed artery. This area of the heart muscle then dies.

The cardiac cycle

The beating of the heart's chambers is called the cardiac cycle. Biologists divide the cardiac cycle into two distinct phases: systole, when the ventricles contract; and diastole, when they relax. At rest, a human heart normally beats about 70 times per minute. You can follow your own cardiac cycle by feeling for your pulse. Gently press two fingers just below the wrist on the underside of your other arm to find your pulse. During systole, the ventricles contract and blood is forced through the arteries just below the skin. The gap between pulses represents diastole, when the ventricles relax.

Speed control

Although the heartbeat is not triggered by the nervous system, two control centers in the medulla oblongata (part of the brain) can influence it. One center sends impulses along accelerator nerves in response to signals from receptors in certain arteries or in the heart itself. The nervous impulses cause the heartbeat to increase in rate and strength. Such an increase is important when extra oxygen is needed by tissues during exercise or times of stress, when the heart must speed up to keep pace with demand. The heart rate can reach 180 beats per minute or more at such times.

If the pressure of the blood gets too high, another set of receptors in the arteries will contact the second control center in the brain. This fires signals along a different pathway, the vagus nerves. The nerve impulses cause the heartbeat to slow down, thus reducing pressure in the arteries.

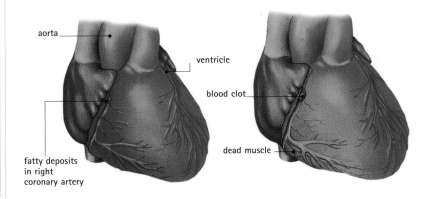

aorta

ventricle

blood clot

dead muscle

fatty deposits
in right
coronary artery

Arteries, veins, and capillaries

CONNECTIONS

COMPARE the circulatory system of a vertebrate such as a *LION* with the method of oxygen and nutrient transfer used by an animal such as a *JELLYFISH*, which does not have a circulatory system.

Like any vertebrate, an insect has a heart to drive fluid around the body, but there the similarity ends. Insects and some other animals have open circulation systems, with fluid washing gently around a central body cavity. In most other animals, including all vertebrates, the nutrient-carrying fluid, blood, moves around the body in tubes called blood vessels. There are three main types of vessels—arteries, veins, and capillaries. Each differs structurally and plays a different role in the path of blood around the body.

Understanding arteries

Arteries carry oxygenated blood away from the heart to the body's tissues and organs. When blood leaves the left ventricle on its journey through the systemic circulation, it enters the largest artery in the body, the aorta. From the aorta, blood is distributed along a branching system of smaller arteries.

The walls of arteries contain many elastic fibers. The elasticity allows the arteries to withstand the high pressure of the blood. The arteries also stretch during systole; each time the left ventricle pumps, the arteries swell a little, storing some of the energy of the beat. The elastic fibers then contract during diastole, when the ventricles relax. This squeezes the blood inside the vessel, pushing it along and keeping the flow relatively smooth.

Blood leaves the arteries along branches of similar but narrower structures called

▼ DEVELOPMENT OF ATHEROSCLEROSIS

(1) The wall of a normal artery is composed of three layers. (2) As atherosclerosis develops, fatty deposits begin to build up in the inner layer of the artery. (3) In extreme cases, the lumen is almost entirely constricted and calcium deposits begin to form.

COMPARATIVE ANATOMY

Vessel structures

Veins and arteries are similar, in that their walls are made up of three main layers. The protective outer layer, the tunica adventitia, is made of collagen and connective tissue. The middle layer, the tunica media, contains muscle and some elastic material. This layer is much thicker in arteries than in veins. The inner layer is the tunica intima, formed by smooth endothelial cells. Arterial tunica intima is lined by an elastic membrane and has elastic fibers on the outside, allowing the walls to snap back when the ventricles relax. Veins lack this elastic material, allowing them to expand and fill with blood.

arterioles. As well as elastic fibers, both arteries and arterioles contain bands of smooth muscle in their walls. This muscle is able to contract, although unlike the stretching of the elastic fibers this is an active process, requiring an energy supply. The muscles allow the diameter of an artery or arteriole to be varied. As the diameter changes, so does the resistance of the vessel to blood flow. In this fashion, the distribution of blood to different parts of the body can be controlled, as can the overall blood pressure.

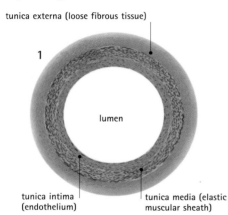

tunica externa (loose fibrous tissue)

1

lumen

tunica intima (endothelium)

tunica media (elastic muscular sheath)

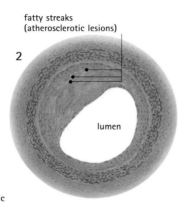

fatty streaks (atherosclerotic lesions)

2

lumen

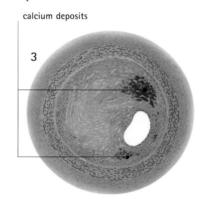

calcium deposits

3

Crucial capillaries

Arterioles open into beds of capillaries. The capillaries form the business end of the circulatory system, where exchange of gases and nutrients takes place between blood and body cells. Capillaries have thin walls, formed by the plasma membranes of a layer of single epithelial cells. Capillaries are also exceedingly fine, so narrow that red blood cells must pass through them in single file. This slows the passage of the blood. The pressure does not increase dramatically inside them, though, because capillaries are incredibly numerous. Their total cross-sectional area is far higher than that of all the other blood vessels in the body, so a bed of capillaries has a far higher capacity for blood than the arterioles that supply it.

From capillary to cell

No body cell is more than a couple of cell widths from a capillary. However, capillaries do not actually connect to the cells that they supply. All capillary walls are leaky, allowing small molecules—such as water, ions, and glucose—to pass through. This passage is driven by the high pressure of the blood, which squeezes these molecules through the capillary wall. The fluid passes into the spaces between the cells.

Red blood cells also release their cargo of oxygen at the capillaries. The release is not driven by blood pressure; instead, the dissociation (temporary breakdown) of waste carbon dioxide released by the cells causes a rise in acidity in the capillary. This causes the hemoglobin (oxygen-carrying molecule) in the red blood cells to release the oxygen to which they are bound.

How exchange takes place

Nutrients and oxygen in the fluid around the cells may enter cells by diffusion—this is the drift of molecules from points of high concentration (in this case, outside the cell) to points of low concentration (inside the cell). Cells may also actively draw in nutrients.

The balance of water in blood and tissue fluid is regulated by a process similar to diffusion, called osmosis. Blood near the end of a capillary is low in water and high in proteins and other molecules that cannot

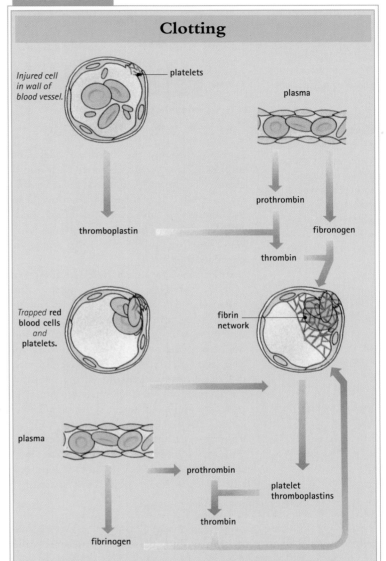

Clotting

Clotting begins when a blood vessel is torn or injured. Platelets arriving at the site of the injury bind to the edges of the wound, forming a patch. Damaged cells release a chemical called thromboplastin into the blood. Thromboplastin converts prothrombin (plasma protein), which is made in the liver, into thrombin. The thrombin acts on the plasma protein fibrinogin, converting it into fibrin. Fibrin is also a protein and is made of long branching fibers that create a weblike structure across the wall of the injured area. This web traps red blood cells and platelets, forming a plug over the wound. The trapped plates release more thromboplastin, which results in the conversion of more fibrinogen to fibrin. The platelets also release contractile proteins, which pull on the edges of the wound, creating a firm clot.

► CAPILLARY

Many different kinds of molecules must be transferred across the walls of capillaries. Protein molecules are engulfed by the walls and then transferred either into surrounding tissues or into the blood plasma. Food molecules, water, and hormones pass through pores. Oxygen and carbon dioxide are exchanged via the walls of the capillaries.

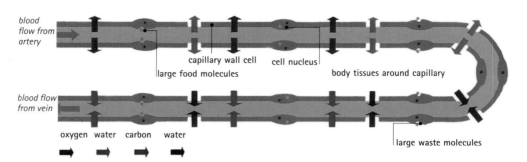

blood flow from artery

capillary wall cell

cell nucleus

large food molecules

body tissues around capillary

blood flow from vein

oxygen   water   carbon   water

large waste molecules

squeeze through the capillary walls. This imbalance causes water to move back into the capillary by osmosis. Osmosis is the diffusion of water molecules across a semipermeable membrane—in this case the capillary walls—from a region where the concentration of water molecules is high to a region where their concentration is low.

Some capillaries have holes in their walls called fenestrations. They allow larger molecules, such as proteins, to pass through. Some have even larger gaps that allow structures as large as blood cells to move in or out. Bone marrow capillaries, for example, must allow newly formed blood cells to enter the bloodstream. Capillaries in the brain have no fenestrations, so only small molecules can cross the wall, although some larger ones, called lipid-soluble molecules, can slip through—these include alcohol. Little can pass from the blood to the brain other than oxygen and glucose. Biologists call this the blood-brain barrier. Most foreign objects cannot cross the

barrier; prions, tiny protein fragments that cause illnesses such as bovine spongiform encephalopathy (BSE) are among the few that are small enough to do so.

The journey back

The fluid that collects around the cells contains water and some small molecules, but no red blood cells. The fluid collects in a separate system of vessels, the lymphatic system. The fluid moves through a series of increasingly large lymphatic vessels, culminating in the thoracic duct. This empties its contents into the superior vena cava, where it reenters the blood.

After leaving a capillary, the blood is at a low pressure. It moves first into venules, then into larger veins. Blood tends to collect in the veins, the walls of which are much more expandable than those of arteries. The veins thus act as a reservoir of blood; at any one time, up to 80 percent of the blood in a human body is in the veins.

COMPARATIVE ANATOMY

Other important functions

Blood vessels have other important functions besides providing a route for blood transport. Some animals have dense networks of capillaries called retia mirabilia. These serve a variety of functions. A rete mirabile in the neck of a giraffe helps keep the blood pressure in the animal's head from rising explosively when the animal lowers its head. The rete mirabile that fills a fish's swim bladder is arranged differently. Blood in adjacent capillaries flows in opposite directions; this is called a countercurrent system. This allows gases to be extracted from the blood efficiently and keeps them from leaking out of the swim bladder. Countercurrent systems involving larger blood vessels may help conserve heat. Bundles of arterioles and small

veins run very close together; warm blood from the core of the body flows in the direction opposite to cool blood from the limbs. Heat is retained as it passes from one set of vessels to the other. This is called a countercurrent heat exchange system. It is very common in the animal kingdom. Baleen whales have a countercurrent heat exchange system in their enormous tongue. Wading birds have this system in their upper legs. Tuna have one in their neck. Unlike most fish, tuna can maintain their body temperature above that of the water around them. They use a countercurrent heat exchange system to conserve heat when they dive into the depths, but blood is rerouted when the fish return to warmer surface waters.

Regulation and control

The heart controls its own beat. Elsewhere in the body, tissues also regulate their blood flow by altering the diameter of the arterioles that supply them. This control is called autoregulation. The body responds to changes in blood flow or content by increasing breathing rate, heart rate, and blood distribution. These changes are coordinated by the endocrine and nervous systems, which influence autoregulatory mechanisms and affect heart output.

Autoregulation in an arteriole depends on the sensitivity of the smooth muscle in the vessel's walls to chemicals in the blood. Low oxygen and high carbon dioxide levels cause the muscle to relax, increasing blood flow. Extra oxygen enters with the blood, and the carbon dioxide is moved swiftly along. Increases in the levels of other products of energy production in the cells, such as lactate, also lead to a faster flow rate.

The muscles in the walls of arteries and arterioles also respond to messages from the endocrine and nervous systems. Sympathetic

adrenal gland

blood under low pressure

kidney

retention of salt in kidney tubules

raised blood pressure

bladder

constriction of arterioles and raised blood pressure

● renin
● angiotensin
● aldosteron

IN FOCUS

Detecting stretch

Information about changes in overall blood pressure is provided by two sets of important detectors. These are stretch receptors, located in the walls of the carotid arteries, which pass through the neck on their way to the brain, and in the aorta. If blood pressure is high, the artery walls are stretched farther than usual. The stretch receptors send messages to the medulla in the brain. The medulla sends nervous signals to the heart, slowing it, and to arterioles in the tissues, which widen to slow the flow. The opposite happens when the pressure is low, except that the stretch receptors also contact the hypothalamus (another part of the brain). This releases a hormone, vasopressin, which also drives arteriole constriction.

nerves in the walls of these vessels in skeletal muscle release acetylcholine; this makes the muscles relax to increase blood flow during exercise. Nerves in vessels of other muscle types release norepinephrine, which makes the muscle contract, reducing blood flow.

A similar hormone, epinephrine, also makes the vessel wall muscles contract. Epinephrine is released from the adrenal glands when a person is suddenly startled or afraid, causing the "fight or flight" response. The hormone minimizes blood flow to nonessential parts of the body, such as the gut. This response increases the overall blood pressure and blood flow to organs, such as the heart and brain, that are important for successful self-defense or escape.

JAMES MARTIN

▲ BLOOD PRESSURE

When blood pressure is low, the kidneys secrete a hormone called renin, which constricts arteries and thus raises blood pressure. At the same time, the adrenal glands produce aldosterone, which causes salt to be retained in the blood. This also raises blood pressure.

FURTHER READING AND RESEARCH
Unglaub Silverthorn, Dee. 2003. *Human Physiology*. Benjamin Cummings: San Francisco, CA.

Digestive and excretory systems

Digestion includes all the ways that Earth's diverse life-forms break down the food they eat into useful and vital products such as proteins, vitamins, minerals, sugars, and starches. The products of digestion fuel life processes, store energy, maintain health, fight disease, and provide the basic building blocks of life.

Plants, seaweeds, single-celled algae such as those in plankton, and some bacteria make their own food from chemicals around them. Plants use the energy of sunlight, gases in the air, and water to make sugars, so they do not need to digest food. Neither do many types of bacteria. Nitrogen-fixing bacteria combine nitrogen with hydrogen, using the energy released by the reaction to produce sugars. Nearly every other type of life-form, however, must eat and digest food to live.

Digesting food creates waste materials. Some products of digestion are harmful toxins; others are of no use to the organism. These wastes are dealt with and removed from the body by excretory systems.

Animal digestion

The complexity of animal digestive systems varies greatly. Specialized systems may be devoted to digesting certain foods, such as the many-chambered tracts of plant eaters. By contrast, gutless tapeworms simply absorb another animal's food. In vertebrates, the digestive system includes all the structures and organs along which food passes through the body.

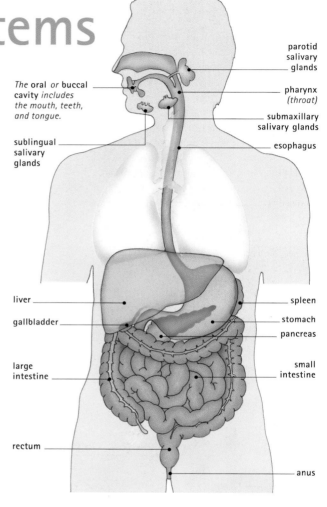

The oral or buccal cavity includes the mouth, teeth, and tongue.

parotid salivary glands
pharynx (throat)
submaxillary salivary glands
esophagus
sublingual salivary glands
liver
gallbladder
large intestine
rectum
spleen
stomach
pancreas
small intestine
anus

SYSTEM HIGHLIGHTS

MOUTH, TEETH, AND TONGUE The mouth, teeth, and tongue begin the process of breaking down food. *See pages 47–49.*

THROAT TO STOMACH The stomach is a stretchy bag containing acids and digestive juices that aid digestion. *See pages 50–53.*

INTESTINES Molecules of digested food, such as sugars and proteins, are absorbed into the bloodstream across the walls of the intestines. *See pages 54–57.*

LIVER, PANCREAS, AND GALLBLADDER The liver cleanses the blood, converting toxins into less harmful products. *See page 58.*

EXCRETORY SYSTEM The kidneys or kidney-like structures filter waste products. *See pages 59–61.*

▲ DIGESTIVE SYSTEM
Human
The digestive system of all vertebrates, including humans, comprises the digestive tract and digestive glands. The digestive tract is a tubular passageway that extends from mouth to anus (or cloaca). Digestive glands line the walls of the tract. They include the pancreas, salivary glands, and the liver. All secrete enzymes that help digest food.

CONNECTIONS

COMPARE how a *MUSHROOM* absorbs its food from organic matter with how plants make food and with how animals digest food.

COMPARE the ways in which *BACTERIA* and a single-celled algae such as a *DIATOM* feed.

Mouth, teeth, and tongue

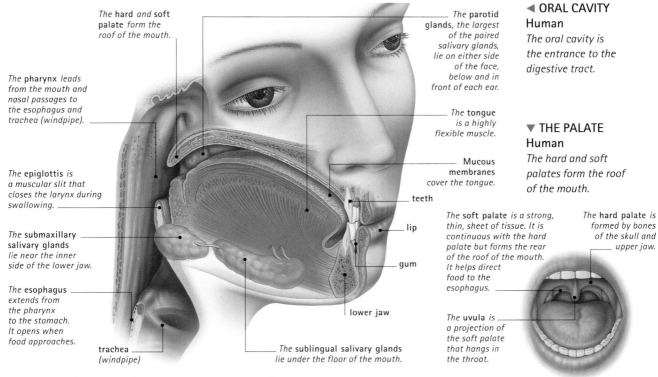

The **hard** *and* **soft palate** *form the roof of the mouth.*

The **pharynx** *leads from the mouth and nasal passages to the esophagus and trachea (windpipe).*

The **epiglottis** *is a muscular slit that closes the larynx during swallowing.*

The **submaxillary salivary glands** *lie near the inner side of the lower jaw.*

The **esophagus** *extends from the pharynx to the stomach. It opens when food approaches.*

trachea (windpipe)

The **parotid glands**, *the largest of the paired salivary glands, lie on either side of the face, below and in front of each ear.*

The **tongue** *is a highly flexible muscle.*

Mucous membranes *cover the tongue.*

teeth

lip

gum

lower jaw

The **sublingual salivary glands** *lie under the floor of the mouth.*

◀ **ORAL CAVITY**
Human
The oral cavity is the entrance to the digestive tract.

▼ **THE PALATE**
Human
The hard and soft palates form the roof of the mouth.

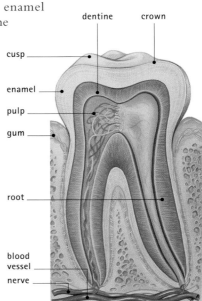

The **soft palate** *is a strong, thin, sheet of tissue. It is continuous with the hard palate but forms the rear of the roof of the mouth. It helps direct food to the esophagus.*

The **hard palate** *is formed by bones of the skull and upper jaw.*

The **uvula** *is a projection of the soft palate that hangs in the throat.*

The vertebrate digestive tract starts with the mouth, teeth, and tongue. These make up the oral, or buccal, cavity. The oral cavity's main task is to prepare food for its journey through the body. In mammals, fleshy lips also allow young to suckle, since they enable an airtight seal, and the mouth, teeth, and tongue are often vital for communicating.

The teeth, lips, inner cheeks, hard palate, and tongue of a mammal all help masticate (chew) food and form it into a rounded mass, or bolus, for swallowing. Only mammals chew. Other animals use their oral structures to catch, hold, or tear up prey. Mastication breaks the food into smaller pieces and mixes it with saliva, which moistens and lubricates the bolus for its journey down the throat. These tasks are mechanical. Saliva also starts the process of chemical digestion. It begins to break down food. The saliva of some mammals, including humans, contains a starch-digesting enzyme, amylase. However, many meat eaters, such as cats and dogs, have no amylase in their saliva, since their diet contains very little starch.

Teeth

A few invertebrates have structures that function as teeth, but only vertebrates have true teeth made from minerals like enamel (in mammals) and dentine. The crown is the visible part above the gum; the root is the unseen part, attached to the jaw by fibers.

Teeth vary a great deal, and their differences are among the criteria used by biologists to help them classify different groups of animals. Many vertebrates have a mouthful

▶ **MOLAR TOOTH**
Mammal
The crown of a mammal's tooth has an outer layer of enamel, the hardest tissue in the body; softer cementum covers the root. Bonelike dentine forms much of the tooth's interior. Blood vessels and nervous tissue make up the innermost pulp.

dentine   crown

cusp

enamel

pulp

gum

root

blood vessel

nerve

of similar-looking teeth. Sharks and other predators, for example, tend to have pointed teeth that can pierce and hold prey. Many fish have fused teeth perfect for scraping food from surfaces. Snakes have backward-curving teeth that grip prey. Turtles and birds have no teeth. Mammals have specialized teeth such as incisors, canines, and molars. The tusks of elephants and walruses are teeth, too. Nearly all vertebrates are born with one set of teeth that are replaced as they wear out.

Most mammals have only two sets of teeth: primary (or first, baby, or milk) teeth, which are replaced by permanent, or adult, teeth. Primary teeth tend to be smaller and more prone to wear than adult teeth. With a few exceptions, adult teeth are not replaced and no longer grow once in place. The incisors of lagomorphs (hares and rabbits) and rodents are worn down by gnawing; these teeth continue to grow from the root throughout the animal's life. An elephant's molars emerge in stages, pushing older molars to the front of the tooth row. Elephants' molars are replaced from behind until quite late in the elephant's life when all the molars have emerged.

CLOSE-UP

Incisors, canines, and molars

Mammals have four main types of teeth: incisors at the front of the mouth, canines on either side of the incisors, premolars along the side of the mouth, and molars at the back. The number and shape of each type vary according to species. Incisors are chisel-shaped and used for cutting and clipping food. Rodents and lagomorphs have long incisors that can gnaw tough plant materials. Canines are pointed teeth that meat eaters use for puncturing, holding, and tearing food. Premolars and molars are large teeth with ridged or peaked surfaces called cusps. They crush and grind food. The fourth upper premolars and first lower molars of carnivores, such as bears, cats, and dogs, are called carnassials. They are used for slicing into the flesh of prey.

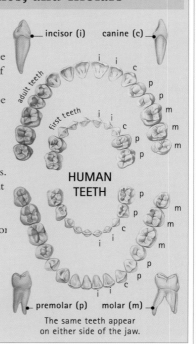

HUMAN TEETH

The same teeth appear on either side of the jaw.

COMPARATIVE ANATOMY

Carnivore, omnivore, herbivore, or bloodsucker?

An animal's teeth meet the requirements of its diet. Carnivores (meat eaters) need sharp teeth that can pierce skin and hold onto prey. The teeth of some carnivores, including snakes and many fish, curve or point backward. Such teeth act as a cage to hold prey in the mouth and direct it toward the throat. Herbivores (plant eaters) need broad, crushing teeth to chew tough leaves and stems. They also need teeth such as incisors to clip leaves from trees or to pull up grass stalks. Omnivores need various types of teeth to deal with their mixed diet.

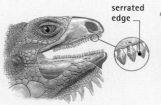

An iguana has peglike teeth with crowns that form a leaf shape. The edges are serrated (jagged) for cutting plants and piercing small prey.

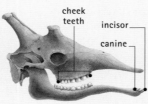

A giraffe has no upper incisors but uses its long tongue and upper lip to pull leaves from trees. Cheek teeth have crescent-shape cusps for grinding.

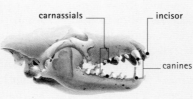

A wolf has large, pointed canines, upper incisors, and carnassial cheek teeth—all perfect for gripping, tearing, and slicing living animal prey.

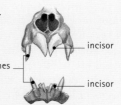

A vampire bat's sharp canines and incisors slice skin. Cheek teeth have few crushing surfaces, since the bat has a liquid diet of vertebrate blood.

IN FOCUS

Spiders have two pairs of feeding appendages. The chelicerae bear fangs that deliver venom. The sensitive pedipalps are used to manipulate food.

Those of males are also used for sperm transfer. The pedipalps often bear enlarged lobes called maxillae, which are used to crush and grind food.

2. The spider regurgitates fluids onto the prey. The fluids contain enzymes that begin to digest the prey before it enters the mouth.

1. A spider buries its fangs in its prey, injecting paralyzing venom. The chelicerae move either forward and down, or together sideways.

3. The spider sucks its food into its mouth. Some spiders rip larger prey into shreds with epidermal teeth on their chelicerae before feeding.

Not all animals, or even all vertebrates, have true teeth. Epidermal teeth are hard, horny projections that function as teeth. Tadpoles, adult platypuses, and jawless fish such as lampreys all have epidermal teeth.

Invertebrate mouthparts

The oral structures of invertebrates vary greatly. Mollusks, such as octopuses, snails, and giant clams, have a hard, tonguelike radula that they use to grind down prey. Corals, hydras, and sea anemones have an opening to their digestive cavity that serves as both mouth and anus. Flatworms, are similar, and also have a series of excretory pores dotted around their body through which wastes are voided. Other invertebrates have a complete digestive tract—that is, a tube with two openings: the mouth and the anus.

Arthropods are invertebrates with jointed limbs and a hard outer exoskeleton. The group includes insects, crustaceans (such as crabs, shrimp, and lobsters), and spiders. Many arthropods have paired, jointed, and limblike appendages or mouthparts called, for example, mandibles, maxillae, chelicerae, and maxillipeds. These appendages are mechanical digesters. They catch, filter, hold, snip, tear up, and generally manipulate food before passing it to the mouth. The form, arrangement, and numbers of each appendage vary immensely between different groups. In lobsters and crayfish the first legs act as mouthparts, too. Either the left or the right claw is massive and has epidermal teeth for crushing hard-shelled prey such as snails and clams.

Filter feeders

Not all animals need to break up large food items. Many animals filter tiny organisms from their watery homes and swallow them whole. Arthropod filter feeders often use appendages (such as maxillae) to filter food from the water. Small aquatic crustaceans called copepods have tiny hairlike setae on their maxillae. Like paddles, these direct food particles to the mouth without touching them. Rorqual whales filter small animals from the water with thin plates of baleen that hang from their upper jaw.

▼ **BEAK AND RADULA**
Common octopus
An octopus's mouth has a strong sharp beak for breaking into hard-shelled prey, and a tonguelike radula covered in hard spikes that grind down prey.

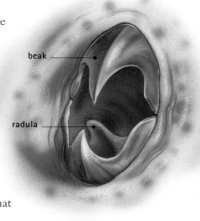

beak

radula

49

Throat to stomach

A vertebrate's foregut consists of the pharynx (throat), esophagus (tube joining pharynx to stomach), and the stomach. Invertebrate foreguts do not always include the stomach, which may be part of the midgut.

The pharynx and swallowing

The pharynx performs roles in the respiratory and digestive systems. In most bony fish it includes the gills, which are located in a series of slits leading from the pharynx to the exterior. In many terrestrial vertebrates the pharynx directs air into the trachea (windpipe) and food into the esophagus. The key digestion-related role of the pharynx is swallowing. Once begun, swallowing is a reflex that cannot be stopped. Most vertebrates bolt their food, swallowing it whole, and the esophagus expands to suit. Seabirds toss food to the back of their throat. In most mammals the back and sides of the tongue expand against the soft palate, forcing food into the esophagus. The epiglottis (a muscular slit) briefly closes the upper part of the trachea, the larynx, which contains the voice box. This prevents food from entering the trachea. In most mammals, the epiglottis forms a seal with the soft palate that keeps the food and air passages separate but open. In adult humans, however, the epiglottis and larynx are much lower, enabling speech. Therefore, the passages are kept separate, but only the esophagus is open during swallowing. That is why adult humans cannot breathe when they swallow. Infant humans can breathe and suckle, however, since their larynx has not yet descended, and the epiglottis–soft palate seal works as it does in other mammals.

The esophagus

The esophagus is a slender, muscular tube that expands around food. Cells in the lining secrete thick, sticky mucus, which greases the passage of food to the stomach. The esophageal lining of animals that eat rough or scratchy foods contains keratin, the key protein in hair, fur, and fingernails. Keratin hardens and protects the lining against damage. Very few esophagi secrete chemicals (enzymes) that digest food.

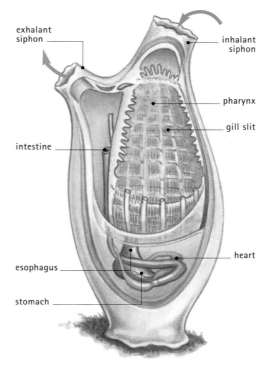

exhalant siphon

inhalant siphon

pharynx

gill slit

intestine

esophagus

stomach

heart

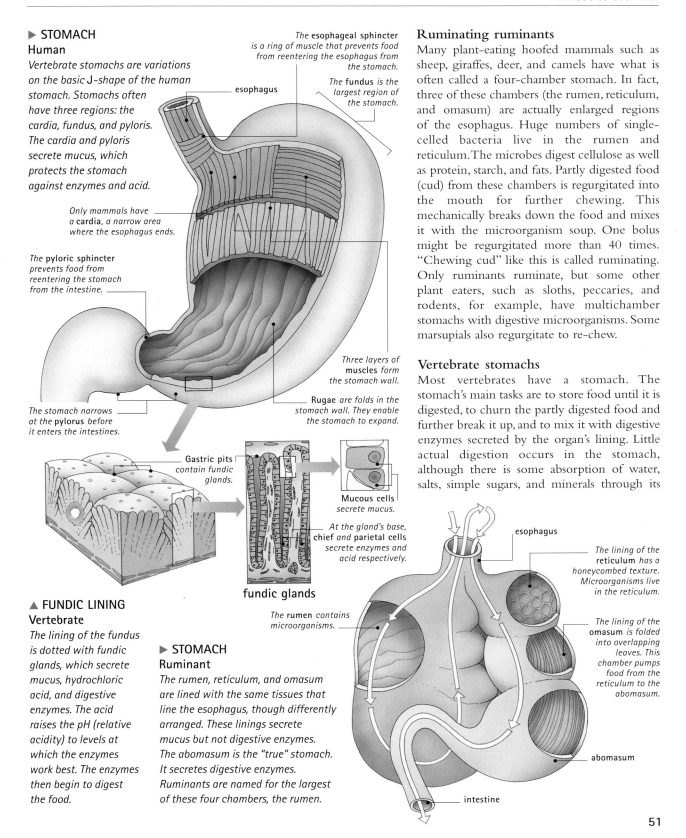

▶ STOMACH
Human

Vertebrate stomachs are variations on the basic J-shape of the human stomach. Stomachs often have three regions: the cardia, fundus, and pyloris. The cardia and pyloris secrete mucus, which protects the stomach against enzymes and acid.

The **esophageal sphincter** is a ring of muscle that prevents food from reentering the esophagus from the stomach.

esophagus

The **fundus** is the largest region of the stomach.

Only mammals have a **cardia**, a narrow area where the esophagus ends.

The **pyloric sphincter** prevents food from reentering the stomach from the intestine.

Three layers of **muscles** form the stomach wall.

Rugae are folds in the stomach wall. They enable the stomach to expand.

The stomach narrows at the **pylorus** before it enters the intestines.

Gastric pits contain fundic glands.

Mucous cells secrete mucus.

At the gland's base, **chief** and **parietal cells** secrete enzymes and acid respectively.

fundic glands

▲ FUNDIC LINING
Vertebrate

The lining of the fundus is dotted with fundic glands, which secrete mucus, hydrochloric acid, and digestive enzymes. The acid raises the pH (relative acidity) to levels at which the enzymes work best. The enzymes then begin to digest the food.

▶ STOMACH
Ruminant

The rumen, reticulum, and omasum are lined with the same tissues that line the esophagus, though differently arranged. These linings secrete mucus but not digestive enzymes. The abomasum is the "true" stomach. It secretes digestive enzymes. Ruminants are named for the largest of these four chambers, the rumen.

Ruminating ruminants

Many plant-eating hoofed mammals such as sheep, giraffes, deer, and camels have what is often called a four-chamber stomach. In fact, three of these chambers (the rumen, reticulum, and omasum) are actually enlarged regions of the esophagus. Huge numbers of single-celled bacteria live in the rumen and reticulum. The microbes digest cellulose as well as protein, starch, and fats. Partly digested food (cud) from these chambers is regurgitated into the mouth for further chewing. This mechanically breaks down the food and mixes it with the microorganism soup. One bolus might be regurgitated more than 40 times. "Chewing cud" like this is called ruminating. Only ruminants ruminate, but some other plant eaters, such as sloths, peccaries, and rodents, for example, have multichamber stomachs with digestive microorganisms. Some marsupials also regurgitate to re-chew.

Vertebrate stomachs

Most vertebrates have a stomach. The stomach's main tasks are to store food until it is digested, to churn the partly digested food and further break it up, and to mix it with digestive enzymes secreted by the organ's lining. Little actual digestion occurs in the stomach, although there is some absorption of water, salts, simple sugars, and minerals through its

esophagus

The lining of the **reticulum** has a honeycombed texture. Microorganisms live in the reticulum.

The lining of the **omasum** is folded into overlapping leaves. This chamber pumps food from the reticulum to the abomasum.

The **rumen** contains microorganisms.

abomasum

intestine

Peristalsis

From esophagus to anus or cloaca, food is pushed through the digestive system by waves of muscular contractions collectively called peristalsis. This occurs in many animals, including invertebrates with tubular digestive systems. The digestive cavities of some invertebrates that do not perform peristalsis are lined with tiny filaments called cilia.

They beat to drive food through the gut. On occasion, peristalsis works in reverse. In birds that feed on fruits with waxy coatings, reverse peristalsis forces intestinal contents back into the gizzard for further grinding and, also, mixing with secretions from the intestine that help break down the fruits' tough skin.

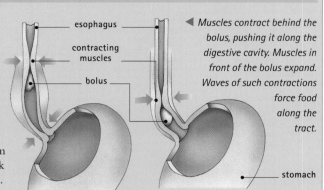

◄ Muscles contract behind the bolus, pushing it along the digestive cavity. Muscles in front of the bolus expand. Waves of such contractions force food along the tract.

walls. Some fish do not have a distinct stomach. A lamprey's diet, for example, consists of blood and flakes of tissue ripped from the body of a host. A lamprey does not need a storage organ; food passes straight from the esophagus to the intestines.

Invertebrate foreguts and midguts

The forms of invertebrate foreguts are as diverse as invertebrates themselves. Some, such as the sandworm or ragworm, do not have a stomach. Instead, esophageal pouches called ceca perform similar tasks. During feeding, the front of the foregut (the pharynx) is turned inside out and everted from the mouth. The inner walls of the pharynx bear teeth and jaws and are lined with tough, shiny chitin, which also lines the outer body. The pharynx retracts when prey is caught, dragging it inside.

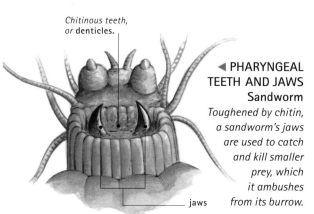

Chitinous teeth, or denticles.

◄ PHARYNGEAL TEETH AND JAWS
Sandworm
Toughened by chitin, a sandworm's jaws are used to catch and kill smaller prey, which it ambushes from its burrow.

jaws

Crops and gizzards

Many birds have an enlarged, thin-walled region of the esophagus called the crop. The crop stores food en route to the stomach and, in some birds, is used to carry food to the young. Also, a gizzard often occurs between the saclike crop and the intestines. The hind part of the stomach, the gizzard has thick muscular walls and contains small stones, or gastroliths, that have been deliberately swallowed by the bird. Gastroliths serve the same function as teeth in other vertebrates, since they crush and grind food, breaking it down mechanically. A few other animals, such as earthworms, alligators, and crocodiles, also have gizzards. Earthworms also have a crop. Some whales and dolphins have an enlarged region of their esophagus that acts like a bird's crop.

▶ DIGESTIVE SYSTEM
Pigeon
A pigeon secretes "milk" from the lining of the crop. This is fed to chicks until around 20 days after hatching.

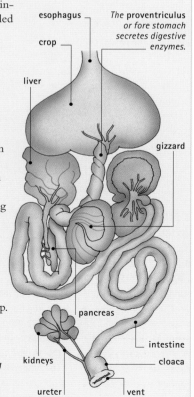

The proventriculus *or fore stomach secretes digestive enzymes.*

esophagus

crop

liver

gizzard

pancreas

intestine

cloaca

kidneys

ureter

vent

Arthropod foreguts and midguts

The foreguts of arthropods, including insects and spiders, also have a chitinous lining. Arthropods have an external skeleton called an exoskeleton. The exoskeleton's outer layer is the chitinous cuticle, which extends into and lines the foregut and hindgut (intestines). The midgut is not lined by cuticle.

Backward-pointing spines or "teeth" might project from the surface of an insect's foregut. An insect's muscular pharynx sucks and swallows, and the esophagus may be enlarged to form a crop. The crop empties into the "stomach," or mesenteron (midgut). Some insects have a muscular gizzard between crop and midgut. Invertebrate gizzards generally contain teeth or grinding plates rather than gastroliths, tiny stones that occur in the gizzards of birds and their ancestors, dinosaurs. Digestion and absorption of the products of the process occur in the insect midgut. The cells that line the midgut have folded surfaces that form fingerlike projections called villi. These are covered by even tinier bumps called microvilli. The projections increase the surface area across which the products of digestion can be absorbed. Midgut pouches called ceca may further increase the surface area.

A crustacean foregut might be a simple tube. However, the foreguts of crabs, lobsters, and other decapods (10-legged forms) contain a structure called the gastric mill. This is a series of hard plates that grind food. Hairlike setae prevent large particles of food from entering the midgut. The midgut often contains one or more diverticula, or pouches, inside which digestion occurs.

Mollusk foreguts and stomachs

Most filter-feeding bivalve mollusks, such as clams and oysters, have gills that pick up oxygen and obtain particles of food from the water. The complex stomach contains one end of a rodlike crystalline style, which rotates and grinds against a hard area of the stomach's wall. As it rotates it dissolves, releasing enzymes that begin to digest the filtered food. In snails and other nonbivalves the foregut includes glands and "teeth" made from tough chitin. These teeth, called the radula, bite, tear, and scrape food materials.

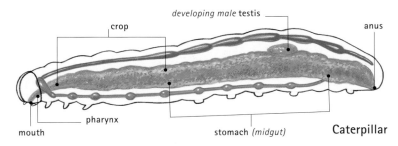

Caterpillar

developing male testis

crop · anus · pharynx · mouth · stomach (midgut)

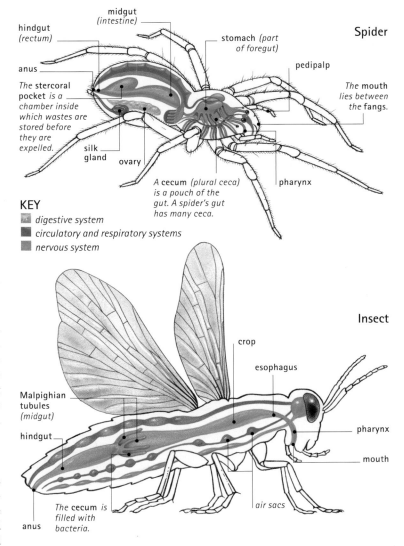

Spider

midgut (intestine) · hindgut (rectum) · stomach (part of foregut) · pedipalp · anus · *The stercoral pocket is a chamber inside which wastes are stored before they are expelled.* · silk gland · ovary · *The mouth lies between the fangs.* · *A cecum (plural ceca) is a pouch of the gut. A spider's gut has many ceca.* · pharynx

KEY

- digestive system
- circulatory and respiratory systems
- nervous system

Insect

crop · esophagus · Malpighian tubules (midgut) · hindgut · pharynx · mouth · anus · *The cecum is filled with bacteria.* · air sacs

▲ DIGESTIVE SYSTEMS
Arthropods

Caterpillars are eating machines, and their large crop, stomach, and intestines reflect this. The pharynx and stomach of a spider are attached to strong muscles. When the muscles contract, the pharynx and stomach dilate. This creates a vacuum that sucks in the spider's liquid food. The ceca of a plant-eating insect might be filled with microorganisms that help digest tough cellulose.

Intestines

In vertebrates, and many invertebrates, too, the intestines are where most chemical digestion takes place. This involves enzymes breaking down food into reusable products such as amino acids, the building blocks of proteins; starchy carbohydrates, which provide energy; and fatty acids. The intestines are also where most products of digestion are finally absorbed into the body. In vertebrates, the products are absorbed through the gut wall, passing into blood vessels in the cavity's lining. They are then carried to where they are needed in the body.

Structure of the digestive tract

From the esophagus to the anus (or cloaca), the digestive cavity of most vertebrates has a common structure. The walls have four layers. The innermost layer is called the mucosa. This includes the epithelium, which lines the cavity, the lamina propria, and the muscularis mucosa. Epithelia vary among species and among locations in the gut. Rodents that eat abrasive things such as insects, seeds, and grasses have epithelia strengthened in places with a tough protein called keratin.

Next to the epithelia lies the lamina propria, a sheet of smooth connective tissue. The muscularis mucosa is a layer of smooth muscle fibers. The muscularis mucosa is so named because it is muscular and because the epithelium contains cells that secrete mucus. Mucus is a thick, sticky fluid that, in the digestive system, aids the passage of food through the tract and protects the cells of the gut wall from the action of digestive enzymes and, in the stomach, acid.

Beneath the mucosa lies the submucosa. This consists of thick but loose connective tissue, and nerves that are part of the autonomic nervous system. This system controls the internal organs without any conscious effort by an animal. Of the sections of the gut, only the esophagus can be contracted at will.

The third of the layers of the mucosa is called the muscularis externa, which has an inner and outer layer of smooth muscle fibers. The inner

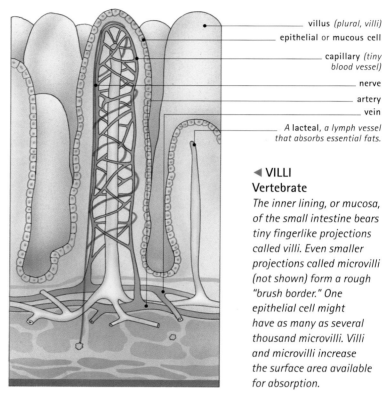

villus *(plural, villi)*
epithelial or mucous cell
capillary *(tiny blood vessel)*
nerve
artery
vein
A lacteal, *a lymph vessel that absorbs essential fats.*

◀ VILLI
Vertebrate
The inner lining, or mucosa, of the small intestine bears tiny fingerlike projections called villi. Even smaller projections called microvilli (not shown) form a rough "brush border." One epithelial cell might have as many as several thousand microvilli. Villi and microvilli increase the surface area available for absorption.

▼ DIGESTIVE TRACT
Vertebrate
The four layers of mucosa, submucosa, muscularis externa, and adventitia (or serosa) are common to the whole digestive tract, from esophagus to anus or cloaca. Glands occur in the mucosa and submucosa.

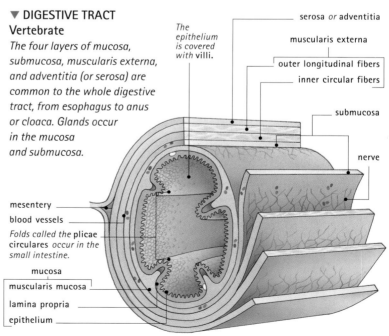

The epithelium is covered with villi.

serosa *or adventitia*
muscularis externa
outer longitudinal fibers
inner circular fibers
submucosa
nerve

mesentery
blood vessels
Folds called the plicae circulares *occur in the small intestine.*
mucosa
muscularis mucosa
lamina propria
epithelium

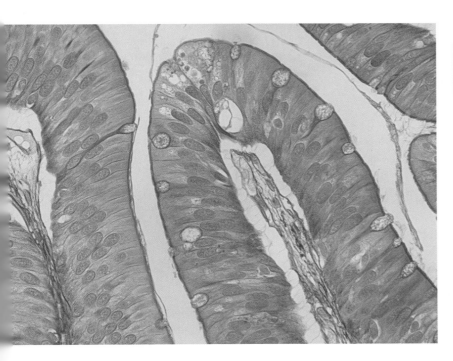

▲ *The gray dots on this cross section from a gut wall are goblet cells. They secrete mucus, a vital fluid that lubricates food as it passes along the gut and helps protect the gut walls from acids and enzymes.*

CONNECTIONS

COMPARE the intestines of a cartilaginous fish such as the *HAMMERHEAD SHARK* with the intestines of a bony fish such as a *SEA HORSE*.

COMPARE the intestines of a carnivore such as a *WOLF* with the intestines of a herbivore such as a *RED DEER* and an omnivore like a *GRIZZLY BEAR*.

layer is circular, while the outer layer runs lengthwise. The fourth, and outer, mucosa layer is the tunica adventitia, which is formed by loose, fibrous connective tissue. In places, the adventitia includes the mesentery. This thin membranous tissue envelops and supports the intestines. As well as an extensive web of blood vessels that carry blood to or from the organs it surrounds, the mesentery contains fatty areas that keep organs warm. Where a mesentery is present, the outer layer of the digestive tract is called the serosa.

Lymphoid tissue occurs throughout the layers of the gut wall, both as nodules and as larger lymph vessels. Lymph is a pale watery fluid containing white blood cells called macrophages and lymphocytes. These cells attack and destroy foreign particles in the body. Fish do not have lymph tissue in their gut wall, but all other vertebrates do.

The small intestines

The intestines of most vertebrates have two main regions: small and large. The terms refer to the diameter of the cavity rather than to its length. The small intestine can be long but is narrower than the large intestine. The duodenum, jejunum, and ileum are the three regions of the small intestine. Soupy chyme

COMPARATIVE ANATOMY

Short or long?

The length of an animal's intestines is often related to diet. Herbivores must digest tough plant material to extract nutrients. A long intestine ensures that food takes a long time to travel through the gut, thus maximizing the time spent on digestion. In herbivores, blind pockets called ceca often occur at the junction of large and small intestines. Ceca further increase the length of the digestive system. Cellulose-digesting microorganisms often live inside ceca.

CARNIVORE   HERBIVORE

Pocketlike or pouchlike extensions called ceca (singular, cecum) further increase the length of many herbivore guts.

Terrestrial carnivores that eat only meat tend to have relatively short guts. Dolphins have relatively long intestines, for carnivores. Sometimes guts are so short, however, that they have evolved specializations to prolong the amount of time food spends in the digestive system. A spiral valve in the digestive cavity of some fish increases the amount of time food spends in their otherwise straight gut by forcing the bolus of food to take a winding path. Lampreys, sharks, lungfish, and sturgeon are among the fish that have spiral valves inside their intestines. Perch and other bony fish generally have no spiral valve but longer, coiled intestines.

Hindgut fermenters

In nonruminant plant-eating mammals such as rabbits and horses, the cecum is a wide branch at the start of the large intestine. Millions of plant-digesting microorganisms live in such a herbivore's cecum. By the time the bacteria have done their job of breaking down the tough plant material, the food has already passed the small intestine. The animal eats partly digested droppings from the cecum to ensure that the small intestine can absorb nutrients such as vitamin B_{12}. Droppings from the cecum are soft and coated in mucus, unlike the dry fully digested pellets. This process is called hindgut fermentation, since the bacteria ferment the food to make sugars as they digest it. Ruminants are foregut fermenters.

▼ *A jackrabbit eats its droppings, a form of feeding called coprophagy.*

(partly digested food) from the stomach enters the duodenum. Duodenal, or Brunner's, glands in the submucosa release chemicals that neutralize the acidic chyme. Other organs release substances into the duodenum. They include the liver, which releases bile via the gall bladder. Bile helps digest fats. The pancreas releases protein-splitting enzymes. The jejunum and ileum are differentiated by the types and structures of their mucosal glands but have no clear demarcation.

The epithelium of the small intestine bears fingerlike villi and microvilli. Villi and microvilli provide a calm, warm spot where chemical digestion can better take place away from the large central cavity. The shape of villi vary in different parts of the small intestine. In the human duodenum, for example, the villi are large, closely packed, and often leaf-shaped. Villi and microvilli also ensure a large surface for absorption, along with folds in the

mucosa called plicae circulares. These folds are present throughout the small intestine except in the first portion, or bulb, of the duodenum and in the lower part of the ileum.

Large intestines

Further absorption occurs in the large intestine, where water especially is drawn back into the body. The large intestine has microvilli but neither plicae circulares nor villi. It joins with the small intestine at a sphincter (ring of muscle) called the ileocolic valve and leads to the anus or cloaca.

The first section of the large intestine is called the cecum. Joined to the cecum is a narrow, blind-ending tube called the appendix. It is a structure unique to humans and apes. In human's distant ancestors it played a role in digestion, but now has no function; it is vestigial. Wombats, civets, rabbits, and many other animals have an appendix-like structure, but these evolved from the cecum independently and are not analogous to the human version.

In mammals the cecum leads to a large hanging portion of the large intestine called the colon. This in turn leads to the final section of the large intestine: the rectum. The rectum narrows to form an anal canal. The expulsion of waste material from the anal canal is controlled, consciously, by a sphincter.

▶ *Tapeworms have no gut. They do not need a digestive cavity, since they are parasites that live inside the guts of vertebrate hosts. A tapeworm absorbs the products of its hosts' digestive system across its skin. This tapeworm lived inside a sheep.*

COMPARE the gut of an insect such as an *ANT* with the intestines of a crustacean such as a *CRAB* or *LOBSTER*, a mollusk such as a *GIANT CLAM*, or a cnidarian such as a *JELLYFISH*.

CONNECTIONS

Invertebrate intestines

Chemical digestion and absorption in invertebrates take place mostly in the intestines. An earthworm, for example, has a relatively simple tubular digestive system. Food passes through the esophagus, is then ground up and mixed in the gizzard, and is stored in the crop until digestion. After that the food is ready to enter the intestines, where digestion is completed and absorption into the body takes place. The intestine's epithelial cells secrete digestive enzymes. As in vertebrates, water is absorbed toward the end of the intestines, and waste exits via the anus.

In insects, microvilli occur in the stomach, which forms part of the midgut, so chemical digestion and absorption do occur in the stomach as well as the hindgut, or intestines. Also, many insects inject saliva into or onto their food before eating, so a considerable amount of chemical digestion takes place before food is even eaten. Similarly, spiders also begin chemically digesting their food before it enters the mouth.

The insect hindgut is generally divided into the pylorus, ileum, and rectum. The pylorus connects the midgut and the hindgut. It sometimes forms a valve. Generally, the ileum is a narrow tube that leads to the rectum. In some insects, the rear of the ileum differs sufficiently to be called a colon. The insect hindgut has a smooth chitinous lining but the layer of tissue (the apical plasma membrane) beneath this cuticle is extensively folded. Like vertebrate villi, these folds increase the surface area available for digestion and absorption.

Like herbivorous vertebrates, the intestines of invertebrates often contain colonies of microorganisms. While some insects have colonies in midgut ceca, others have them in their ileum. Larval (young) scarab beetles have an enlarged microorganism-containing ileum called a fermentation chamber, and termites have one called a paunch. The microorganisms enable the insects to digest tough plant material, even wood.

The insect rectum contains pads that reabsorb water. The rectal chamber may have other functions. Dragonfly nymphs develop in freshwater. Their rectum is lined by gills. They pump water over their rectal gills to enable respiration. They also use the pumping muscles to force water from the chamber, jet-propelling the insect through the water.

CLOSE-UP

Invertebrates without intestines

Many invertebrates do not have a digestive cavity with a separate foregut and hindgut. Hydras are small aquatic invertebrates related to sea anemones and corals. They have a simple canal-like cavity inside their body. The entrance serves as both mouth and anus. The products of digestion absorb directly into body tissues. Corals, sea anemones, and jellyfish have similar guts. Planarian flatworms have a more complex, branching gut. Food is broken down by enzymes, then absorbed by cells lining the digestive cavity and passed to the bloodstream. The branches of the flatworm's gut deliver the products of digestion to the whole body. Wastes exit through the mouth or through excretory pores. Tapeworms have no gut at all. They absorb the products of their host's digestion and so do not need a gut of their own.

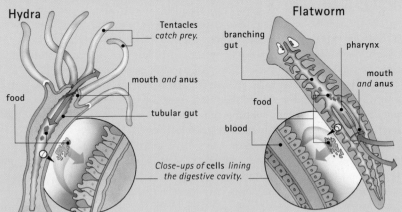

Hydra — Tentacles *catch prey.* — mouth *and* anus — food — tubular gut — *Close-ups of cells lining the digestive cavity.*

Flatworm — branching gut — pharynx — mouth *and* anus — food — blood

Cells that line the digestive cavity of a hydra digest food both externally (by releasing enzymes) and internally by engulfing particles for digestion within the cell.

A flatworm's mouth is on the underside of its body. The pharynx extends out of the mouth like a tube during feeding. Blood carries digested food products to body tissues.

Liver, pancreas, and gallbladder

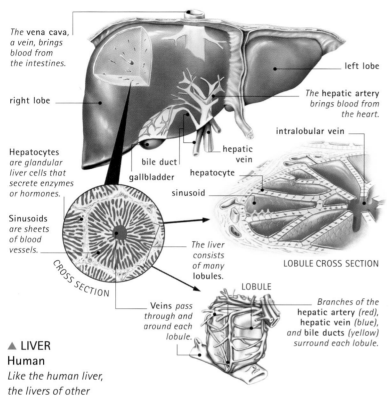

The **vena cava**, a vein, brings blood from the intestines.

right lobe

Hepatocytes *are glandular liver cells that secrete enzymes or hormones.*

Sinusoids *are sheets of blood vessels.*

left lobe

The **hepatic artery** brings blood from the heart.

intralobular vein

hepatic vein

bile duct

gallbladder

hepatocyte

sinusoid

CROSS SECTION

The liver consists of many lobules.

LOBULE CROSS SECTION

LOBULE

Veins *pass through and around each lobule.*

Branches of the **hepatic artery** *(red),* **hepatic vein** *(blue), and* **bile ducts** *(yellow) surround each lobule.*

▲ **LIVER**
Human
Like the human liver, the livers of other vertebrates are housed within the rib cage. The liver's shape depends on the shape of the animal. A snake's liver is long and narrow, for example. Most livers have two lobes, left and right. The microscopic structures of a liver do not vary, however. Sheets of hepatocytes are interspersed with sinusoids (sheets of blood vessels). Venous blood (from veins) carries the products of digestion to the liver, and arterial blood brings blood from the heart.

In vertebrates the liver, gallbladder, and pancreas are important glands of the digestive system. They are called associated glands since they do not lie within the digestive system, unlike the mucus- and enzyme-secreting glands that occur within the walls of the digestive tract. All three glands release chemicals into the duodenum.

Liver and gallbladder
All vertebrates have a liver. It is the largest gland and, after the skin, the largest organ of the body. The liver has several tasks. When an animal is an unborn fetus, the liver makes red blood cells. After birth, the liver destroys old red blood cells. One task the liver always performs is to detoxify (remove poisons from) the blood. Plant eaters, in particular, eat a lot of toxins; many plants store chemicals in their leaves and stems to defend themselves. The liver also produces bile, and stores carbohydrates, proteins, and fats, converting them

into other materials when needed. The liver contains a huge number of blood vessels, more so than nearly every other organ.

The gallbladder is a small organ that stores bile and releases it into the duodenum when needed. Bile emulisifes fats, breaking them into small globules that provide a greater surface area on which digestive chemicals can act. Most, but not all, vertebrates have a gallbladder. Lampreys, hagfish, most birds, and a few mammals do not have one.

The pancreas
The pancreas is an organ that makes and releases pancreatic juice, which largely comprises the protein-splitting enzyme trypsin. Other enzymes in the juice break down carbohydrates and fats. The pancreas also produces the hormones insulin and glucagon, which are proteins that regulate the levels of glucose (a sugar) in the blood. All vertebrates have a pancreas, though it is not always present as a single, discrete organ. The pancreas of lampreys and hagfish are gland-containing tissues spread throughout the intestinal submucosa and on the liver. Tetrapods (reptiles, amphibians, birds, and mammals) have a single discrete organ, as in humans.

Invertebrate equivalents
Invertebrates have neither a liver, a gallbladder, nor a pancreas. Some, however, have tissues that perform similar tasks. Lancelets are fishlike invertebrates that are close relatives of vertebrates. They have a pouch in their gut, the hepatic cecum, which develops in a similar position to the liver of a vertebrate. Veins bring blood carrying the products of digestion to the organ. Its tasks differ from those of a true liver, however, since it is where digestive enzymes are made and food is absorbed.

Earthworms have chlorogogen, a yellowish tissue circling the gut. It functions like a liver, storing glucose as glycogen and releasing it when needed and breaking down toxins. It also acts as a fat store and makes hemoglobin (the oxygen-carrying particle in the blood).

Excretory system

All life-forms must rid their body of toxins and wastes. In animals this is performed by a variety of excretory systems, including the kidneys in vertebrates. The kidneys produce urine and share ducts with the reproductive system. For that reason, both the urinary and reproductive systems are sometimes termed the urogenital system.

Excretion not only expels wastes. It also ensures that the body maintains the correct balance of vital chemicals—a regulatory function termed homeostasis. Aquatic life-forms, for example, balance the levels of salts in their body to prevent the excessive osmosis (diffusion) of water out of or into the body. Diffusion is the tendency for particles to move from a region of high concentration to a region of low concentration. Osmosis is the diffusion of water across a partly porous membrane, such as a cell membrane or nonwaterproof skin. Creatures that live in hot or dry places must balance salt levels to conserve water. Kidneys are vital water-saving organs for terrestrial vertebrates.

Defecation and urination

The digestive system voids feces at the anus or cloaca. Feces mostly comprises the undigested remains of food, along with other wastes such

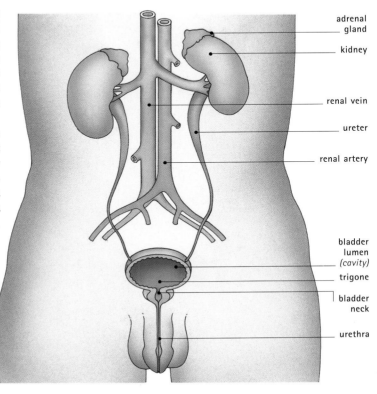

▲ URINARY SYSTEM
Human male

In humans, other mammals, and many other vertebrates, the kidneys produce urine. The ureters, bladder, and urethra *are vital for the passage, storage, and excretion of urine. The bladder is a hollow organ of varying capacity. It has a muscular coat that empties the organ when it contracts.*

adrenal gland
kidney
renal vein
ureter
renal artery
bladder lumen (cavity)
trigone
bladder neck
urethra

COMPARATIVE ANATOMY

Mammalian cloacae

A cloaca is a common chamber through which feces and urine pass and which also contains the genital opening. Cloacae occur in reptiles, birds, and amphibians, but are found in just a few types of mammals. These are the monotreme, or egg-laying mammals, which include the platypus and two species of echidnas.

Female monotreme

kidney
ovary
colon
bladder
urethra

Reproductive, excretory, and digestive systems merge into a cloaca, with one exit.

Female marsupial

kidney
ovary
bladder
urethra
colon

Excretory and reproductive systems fuse. The digestive system has a separate exit.

Female placental

kidney
ovary
urethra
bladder
colon

Each system has a separate exit to the outside.

► KIDNEY
Mammal

The kidneys of birds and reptiles are similar to those of mammals. All have tiny coiled tubes called nephrons, which filter waste from the blood and reabsorb water and nutrients. In mammals the final result is urine, which is expelled via the bladder.

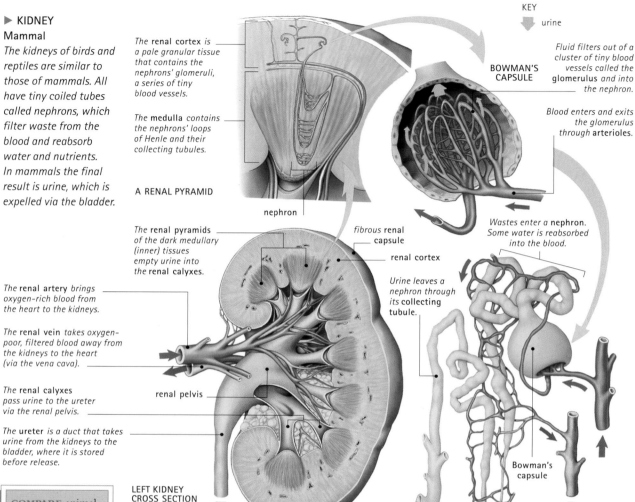

The **renal cortex** is a pale granular tissue that contains the nephrons' glomeruli, a series of tiny blood vessels.

The **medulla** contains the nephrons' loops of Henle and their collecting tubules.

A RENAL PYRAMID

nephron

The **renal pyramids** of the dark medullary (inner) tissues empty urine into the **renal calyxes**.

The **renal artery** brings oxygen-rich blood from the heart to the kidneys.

The **renal vein** takes oxygen-poor, filtered blood away from the kidneys to the heart (via the vena cava).

The **renal calyxes** pass urine to the ureter via the renal pelvis.

The **ureter** is a duct that takes urine from the kidneys to the bladder, where it is stored before release.

renal pelvis

LEFT KIDNEY CROSS SECTION

BOWMAN'S CAPSULE

Fluid filters out of a cluster of tiny blood vessels called the **glomerulus** and into the nephron.

Blood enters and exits the glomerulus through **arterioles**.

fibrous renal capsule

renal cortex

Wastes enter a **nephron**. Some water is reabsorbed into the blood.

Urine leaves a nephron through its **collecting tubule**.

Bowman's capsule

loop of Henle

NEPHRON

COMPARE animal excretion to that of a single-celled protist such as a plantlike **DIATOM** or an animal-like **AMOEBA**. The metabolic wastes of these protists simply diffuse out of the cell, exit via holes called anal pores, or are retained inside vacuoles.

COMPARE how animals produce urine and feces to how **BACTERIA** convert these into useful compounds.

as the remnants of broken-down red blood cells. This process is called defecation. The urinary system excretes nitrogen-based wastes in a process called excretion. These wastes are made from the by-products of vital cellular processes collectively termed metabolism. Ammonia is a by-product of protein break down. This gas is toxic to most animals. In small aquatic animals it dissolves in water and diffuses out of the animal through the skin. Mammals convert ammonia into a chemical called urea, which dissolves into water and is passed out of the body in urine. Birds, reptiles, and terrestrial insects convert metabolic wastes into uric acid. This does not easily dissolve in water, so less water is needed to excrete it. Uric acid is excreted in the form of a paste.

Vertebrate excretion

Many vertebrates, including all mammals, have a pair of kidneys or kidneylike organs. The mammalian kidney regulates osmosis, balances chemicals, and produces urine. These functions are performed by separate organs in many non-mammals. The gills of many fish, for example, remove nitrogen compounds from the body. Many seabirds have glands that excrete salt, enabling them to drink seawater. Birds and reptiles do not have a bladder. Wastes are converted into uric acid. This is emptied into the end of the digestive tract and excreted via the cloaca with feces. Amphibians have a large bladder to store water in when they are on land. Also, the amphibian kidney increases or decreases its rate of filtration depending on whether the animal is in or out of the water and needs to conserve or expel water.

Amphibians and freshwater fish expel a lot of urine. This is because water constantly moves into the body through osmosis (since their body is saltier than the surroundings). Marine fish constantly lose water to their saltier surroundings. They must swallow a lot of seawater to counteract this, and conserve as much water form urine as possible. Most marine invertebrates are of similar saltiness to the seawater around them, so water neither moves in nor out.

Invertebrate excretion

There are four main types of invertebrate excretory organs—nephridia, renal glands, coxal glands, and Malpighian tubules. Mollusks have one or two kidneylike organs called renal glands, which filter metabolic wastes from body fluids. They are excreted as ammonia, ammonium chloride (in octopuses), or uric acid (in slugs and snails).

Aquatic arthropods such as crustaceans have paired coxal glands, which open at the bases of limbs or antennae. The gland is a coiled tube that empties urine into a bladder or a duct to the bladder. The tube begins as a small sac called the celomic sac, which filters the blood in a similar way to the glomerulus and Bowman's capsule of the vertebrate kidney.

Cellular elimination

Plants must protect themselves against toxins, too. Many plants contain poisons that deter animals from eating them. A plant cell typically contains a membrane-bound, fluid-filled space called a vacuole. Plants isolate poisons within their vacuoles. Coffee plants store caffeine within their vacuoles; tobacco plants store nicotine in theirs. Microorganisms expel poisons by enclosing them in vesicles. The vesicles merge with the cell membrane then empty their contents outside.

BRIDGET GILES

FURTHER READING AND RESEARCH
Arnold, Nick. 1999. *Horrible Science: Disgusting Digestion*. Scholastic: Danbury, CT.

Nephridia and Malpighian tubules

Earthworms have excretory organs called nephridia. Each nephridium is a long, fine tubule that opens in the body cavity and leads to the outside, dumping watery wastes onto the skin. Flatworms have excretory organs called protonephridia. They lead to clusters of cells called flame cells.

Insects use organs called Malpighian tubules to excrete wastes. Some insects have one pair, others have more than a hundred. Malpighian tubules start in the body cavity, where they are bathed in hemolymph (equivalent to blood). Wastes are pumped into the tubule from the blood. The tubules do not open onto the exterior but instead empty into the digestive tract, at the junction of the midgut and hindgut. The urine produced by the tubules passes through the rectum, where water and other useful chemicals are reabsorbed.

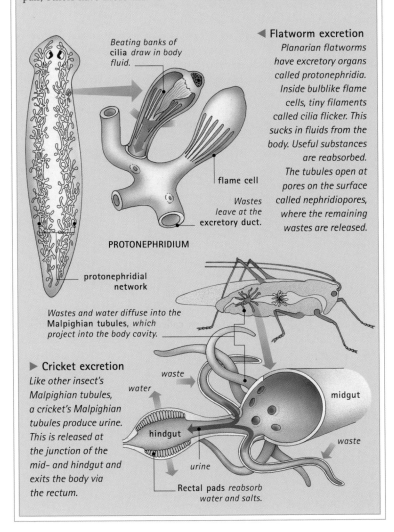

Beating banks of cilia *draw in body fluid.*

flame cell

Wastes leave at the **excretory duct.**

PROTONEPHRIDIUM

protonephridial network

◀ **Flatworm excretion**
Planarian flatworms have excretory organs called protonephridia. Inside bulblike flame cells, tiny filaments called cilia flicker. This sucks in fluids from the body. Useful substances are reabsorbed. The tubules open at pores on the surface called nephridiopores, where the remaining wastes are released.

Wastes and water diffuse into the **Malpighian tubules,** *which project into the body cavity.*

▶ **Cricket excretion**
Like other insect's Malpighian tubules, a cricket's Malpighian tubules produce urine. This is released at the junction of the mid- and hindgut and exits the body via the rectum.

waste

water

waste

hindgut

urine

midgut

waste

Rectal pads *reabsorb water and salts.*

Endocrine and exocrine systems

Animals use nerves to send messages around the body within a fraction of a second, to regulate functions such as pain sensation or muscular activity. However, many life processes take place over much longer time scales. Many of these processes—both in animals and in plants—are controlled by hormones. Hormones are chemical signals that initiate some action in the body. They control the day-to-day functions of the body, such as digestion, reactions to stress, and regulation of sugar levels in the blood. Hormones also regulate longer-term physical changes, such as growth and development.

Hormones are produced by a network of glands called the endocrine system. Hormones act inside the body. Not all secretions from glands are hormones, though. A separate network called the exocrine system releases chemicals to the outside of the body and onto the surfaces of cavities inside it. Sweat and salivary glands, for example, are important parts of the exocrine system.

Nerves, glands, and the endocrine system

All organisms that have a nervous system also have some type of endocrine system. When body conditions need fine-tuning, the nervous system signals the glands or glandlike tissues of the endocrine system. The glands then release the relevant hormones. Most endocrine glands in vertebrates are tightly packed cells that form thin layers called epithelial tissues. Glands are generally small and are well supplied with blood vessels; they rely on blood flow to carry their chemical signals around the body. The glands of fish, amphibians, and reptiles tend to be diffuse, spread throughout a tissue. In mammals and birds, most glands consist of a single clump of secretory cells.

Most endocrine glands have no ducts but are surrounded by other tissues. These ductless glands secrete hormones directly into the spaces between cells, from where they drift into the bloodstream. By contrast, most exocrine glands have ducts that channel their secretions away.

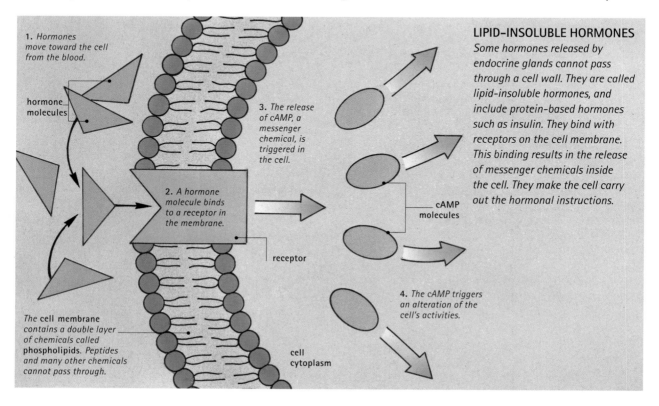

1. *Hormones move toward the cell from the blood.*

hormone molecules

2. *A hormone molecule binds to a receptor in the membrane.*

3. *The release of cAMP, a messenger chemical, is triggered in the cell.*

receptor

The **cell membrane** *contains a double layer of chemicals called* **phospholipids.** *Peptides and many other chemicals cannot pass through.*

cell cytoplasm

cAMP molecules

4. *The cAMP triggers an alteration of the cell's activities.*

LIPID-INSOLUBLE HORMONES

Some hormones released by endocrine glands cannot pass through a cell wall. They are called lipid-insoluble hormones, and include protein-based hormones such as insulin. They bind with receptors on the cell membrane. This binding results in the release of messenger chemicals inside the cell. They make the cell carry out the hormonal instructions.

Vertebrate hormones transmit their messages in three ways. In autocrine signaling, the hormone acts on the cell that secretes it. In paracrine, or local, signaling the hormone acts on nearby cells. In endocrine, or long-distance, signaling, the hormone circulates in the blood and acts on distant cells.

How hormones work

Each hormone can act only on its target cells. They have receptors that can bind only to that hormone. Nontarget cells are also exposed to the hormone. However, they do not have the correct receptor, so they cannot respond to the hormone. Hormones are either lipid- (fat-) soluble or lipid-insoluble. Lipid-soluble hormones can penetrate cell membranes and bind to receptors inside the cell. Lipid-soluble hormones often have long-term effects on the body, controlling growth and development. Lipid-insoluble hormones are not able to enter cells. Instead, they bind to receptors on the cell's outer membrane.

When a hormone binds to its receptor, the cell is induced to perform a specific function. Some receptors do this directly by moving into the cell with the hormone molecule attached. Others activate new chemicals that carry the message into the cell. Either way, a cascade of chemicals is released that carries out the hormone's instructions.

One of the most important messenger chemicals inside a cell is called cyclic adenosine monophosphate, or cAMP. When a hormone such as glucagon or calcitonin binds to a receptor, cAMP is produced in the cell's membrane. The cAMP moves into the cell, where it triggers proteins called enzymes. They drive chemical reactions in the cell, and so control the way the cell behaves.

▲ In contrast to the endocrine system, the exocrine system produces secretions that may act externally. This strawberry poison arrow frog, for example, has glands that secrete a powerful and deadly poison onto the skin's surface. This wards off predators. The poison is produced by exocrine skin glands.

Types of hormones

The hormones produced by mammals fall into four main categories. Amines, such as thyroid hormones and catecholamines, are derived from amino acids. Eicosanoids, such as prostaglandins, are produced from chemicals called fatty acids. These hormones are stored in cell membranes and are released as soon as they are needed.

Most hormones belong to the other two groups: steroids and protein-based hormones. Steroids are lipids (fats) derived from cholesterol. They include estrogen and testosterone, which are made and secreted by glands in the sex organs. These hormones travel in the blood, and are not stored by cells. Steroid hormones are able to pass easily through the outer membranes of cells. This ability allows them to reach their specific receptors inside.

Protein-based hormones range from small peptide hormones composed of as few as three tiny amino acid molecules to giant, complex protein molecules. Insulin and the pituitary hormones are protein-based. Unable to enter cells directly, peptide hormones are stored as granules in glands and are released into the bloodstream in bursts.

SYSTEM HIGHLIGHTS

GROWTH AND DEVELOPMENT Growth from infancy through adulthood is controlled by hormones such as somatotropin. Hormonal activity drives even more dramatic changes in animals such as frogs and insects. *See pages 64–65.*

HOMEOSTASIS The body's internal environment—for example, temperature, salt concentration, and water level—is under strict hormonal control. *See page 66.*

REPRODUCTION Hormones control the development of the sex organs; they also control features such as the release of eggs from the ovaries. *See page 67.*

VERTEBRATE ENDOCRINE SYSTEMS Regulated by the hypothalamus and pituitary gland, vertebrate endocrine systems drive a suite of complex interactions. *See pages 68–71.*

EXOCRINE SYSTEM Exocrine glands release their important secretions onto the body surface or into body cavities. *See pages 72–73.*

Growth and development

▼ HYPOTHALAMUS
AND PITUITARY
Human
*The hypothalamus
secretes chemicals
called neurohormones.
They travel down the
portal vein to the
anterior pituitary,
causing it to release
particular hormones
into the bloodstream.*

In all vertebrates, invertebrates, plants, and even fungi, hormones control crucial body functions, such as growth, development, homeostasis (the regulation of an organism's systems), and reproduction.

Growth hormones

Every organism needs to grow. The regulation of growth by hormones occurs in even the smallest multicellular organisms. Vertebrate growth is controlled by the hypophysis, a master gland that occurs in the brain of most animals. In humans, this corresponds to the pituitary gland, a bean-sized gland. It lies in the brain just below another gland called the hypothalamus. Most animals have a structure called a hypothalamus or a similar gland in the brain that secretes neurosecretory hormones. These work directly on the hypophysis. Some neurosecretory hormones trigger hormone release from the hypophysis; others prevent hormone release.

Hypophysis structure is remarkably similar in most vertebrates. In most tetrapods (four-limbed vertebrates) the hypophysis consists of two sections called the pars intermedia and the pars distalis. These sections correspond to the posterior and anterior pituitary glands in humans. A growth hormone called somatotropin is secreted by the pars distalis on orders from the hypothalamus. In humans, somatotropin is synthesized and released from the anterior pituitary gland.

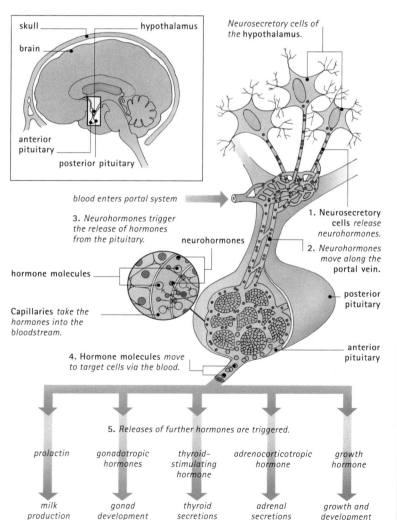

skull
hypothalamus
brain
anterior pituitary
posterior pituitary

Neurosecretory cells of the hypothalamus.

blood enters portal system

3. Neurohormones trigger the release of hormones from the pituitary.

neurohormones

1. Neurosecretory cells *release neurohormones.*

2. *Neurohormones move along the portal vein.*

hormone molecules

posterior pituitary

Capillaries *take the hormones into the bloodstream.*

4. Hormone molecules *move to target cells via the blood.*

anterior pituitary

5. *Releases of further hormones are triggered.*

prolactin — *milk production*

gonadotropic hormones — *gonad development*

thyroid-stimulating hormone — *thyroid secretions*

adrenocorticotropic hormone — *adrenal secretions*

growth hormone — *growth and development*

IN FOCUS

Growth in plants

Plants have growth hormones, too. Auxin is a hormone produced by cells in the growing regions of a plant. Auxin is plentiful at the tips of shoots, where it controls their elongation and upward growth. It is also present in the roots, triggering them to grow down into the soil.

Other plant hormones include gibberellins, which occur on leaf tips and work with auxin to elongate cells during plant growth. In some plants, gibberellins also stimulate the production of enzymes used in metabolism. These enzymes include cytokinins, which regulate cell division and, with auxin, cause cell differentiation (the formation of distinct types of cells) to take place. Other plant hormones include ethylene, which stimulates the ripening of fruit and the germination of seedlings; and abscisic acid, which regulates the falling of leaves in deciduous trees.

Keeping growth in check

Growth cannot occur unhindered. The hormone that induces growth must itself be controlled. In humans, a hormone called somatostatin inhibits growth. Most animals have growth-inhibiting hormones. Bony fish have a gland called a urophysis gland, which is located near the base of the tail. The urophysis produces two hormones (urotensins I and II) that perform the same function as the growth-inhibiting somatostatin in humans.

The hypothalamus produces a thyroid-releasing hormone (TRH); this stimulates the hypophysis to produce thyroid-stimulating hormone (TSH). This in turn acts on the thyroid gland, which is located in the neck in humans. TSH activates the production and secretion into the blood of the thyroid hormones, T3 and T4, both of which are crucial for growth and development. Similar thyroid hormones control growth in cephalopods such as octopuses.

Making adults

Thyroid hormones play a vital role in the metamorphosis (change) of frogs from tadpoles into adults. Adult frogs are insect-eating, tailless, air-breathing, and four-legged; but tadpoles are vegetarian, long-tailed, gilled, and legless. A thyroid hormone, thyroxine, controls metamorphosis in frogs. When it is young, a tadpole's body contains very little thyroxine. When the tadpole is ready to metamorphose, a signal from the brain causes the thyroid gland to begin pumping out thyroxine. The hormone controls the disappearance of the tail and the development of limbs and other adult features. In mammals, thyroxine helps regulate metabolism.

Important insect hormones

Growth in arthropods such as insects involves the shedding of the hard outer skin, or exoskeleton. This process, called molting, is under close hormonal control. The brain releases a peptide hormone, PTTH, that stimulates the prothoracic glands. They lie at the front of the thorax. These glands make and secrete ecdysone, a steroid hormone that controls the shedding of the exoskeleton. Another hormone, juvenile hormone (JH), shuts down production of ecdysone when

IN FOCUS

The importance of metabolism

Growth depends on how quickly food can be converted into energy that can be used to make new cells. The conversion of food into energy is called metabolism. The metabolism of all vertebrates is regulated by hormones. The growth hormone somatotropin enables the metabolism of fats, which provide the body with the energy needed for growth. Somatotropin also stimulates the production of the other growth hormones. Many of these are produced in the liver and control the breakdown of proteins and carbohydrates. In humans, the pancreas and the adrenal cortex, among other glands, produce hormones that control the metabolism of glucose (a type of sugar).

Not all invertebrates have metabolic hormones, but insects do. Beside an insect's brain is a paired structure called the corpora cardiaca. This structure secretes a hormone that controls the metabolism of trehalose, a type of sugar that occurs in insect bodies.

molting is complete; the old exoskeleton has been shed, the animal has swelled a little, and a new, larger exoskeleton covers the body. Ecdysone functions only when the insect is growing. Adult insects do not molt; their prothoracic glands degenerate, and ecdysone production stops. JH is produced by glands called the corpora allata. They lie on either side of the esophagus near the brain.

Insects from several groups undergo the process of complete metamorphosis; a larval insect, such as a caterpillar, changes into an adult (in this case, a butterfly) through an immobile pupal stage. This process is regulated by JH. If lots of JH is present when the insect molts, it will pass to another larval stage. As JH levels drop, the larva molts into a pupa. A complete absence of JH allows the pupa to transform itself into the adult insect.

▼ The process by which tadpoles develop into frogs is called metamorphosis. This change is controlled by the hormone thyroxine, which is produced by the tadpole's thyroid gland.

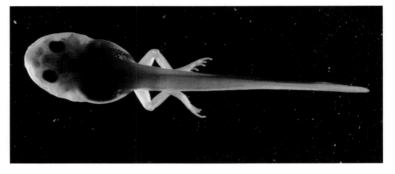

Homeostasis

This diagram shows the way in which thyroid hormone is regulated using a system called a negative feedback loop. When the level of thyroid hormone in the blood is higher than normal, the production of thyroid-stimulating hormone is inhibited. Thus, less thyroid hormone is produced.

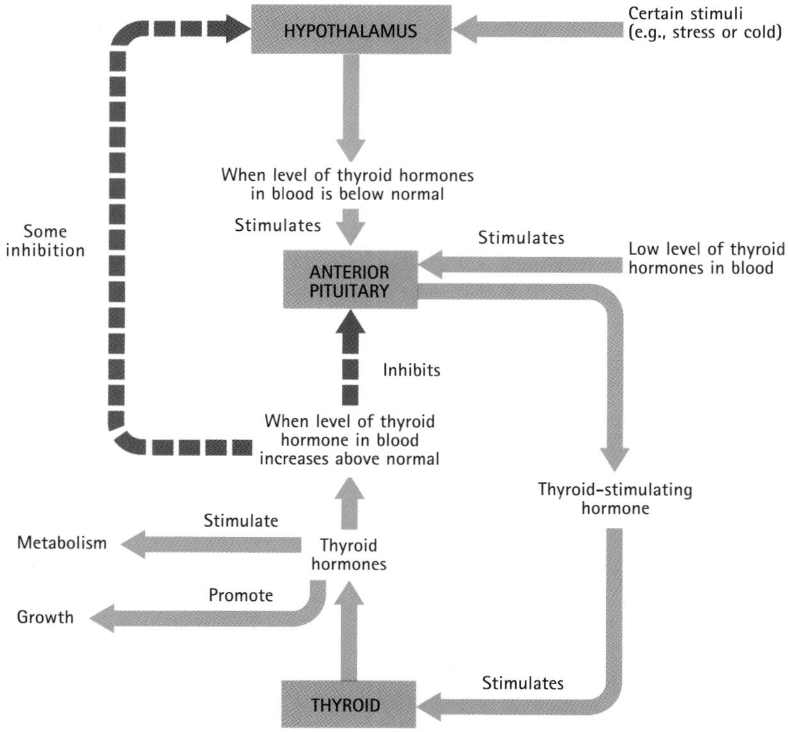

Homeostasis is the regulation of the body's internal environment, features such as salt concentration, water levels, and temperature. Hormones are an important part of homeostasis. For example, they strictly regulate the concentrations of ions (charged particles) in the blood or in the extracellular matrix (the spaces between cells). Calcium, potassium, sodium, and phosphorus ions are the primary targets of this hormonal control system.

Birds, reptiles, amphibians, and fish have paired glands called pharyngeal ultimobranchial glands that produce and secrete the peptide hormone calcitonin. This controls the levels of calcium. Bony fish also have a unique structure called the stannius gland in the kidneys. The stannius gland produces a peptide hormone called hypocalcin, a potent calcium regulator.

Mammals produce calcitonin in C-cells, which lie in the thyroid gland. Calcitonin reduces levels of calcium in the blood in

Water regulators

The posterior pituitary glands of vertebrates produce a crucial peptide called antidiuretic hormone, or ADH. A diuretic speeds up the process of urine production. ADH regulates the amount of water in the body by acting on the kidneys. This allows the body to regulate urine output. Many insects produce an antidiuretic hormone similar to that of vertebrates.

two ways. First, the calcitonin suppresses the action of cells called osteoclasts that digest bone, releasing calcium and phosphorus into the blood. Second, the hormone increases the removal of calcium and phosphorus from the urine by acting on the kidneys.

The parathyroids and adrenals

Most vertebrates (though not fish) have a parathyroid gland, which consists of four pea-sized nodules attached to the back of the thyroid. Parathyroid hormone is the most important controller of concentrations of calcium and phosphorus in the body; levels of these ions must remain within very narrow ranges; otherwise, muscles and the nervous system cease to function correctly. Parathyroid hormone works in the opposite way to calcitonin, with the additional function of increasing the absorption of calcium from the small intestine. Parathyroid hormone is released in response to low levels of calcium and phosphate in the blood. It brings levels of these ions back to within normal range.

Humans have a pair of adrenal glands embedded in a wad of fat in front of the kidneys. One part of each, the adrenal cortex, is crucial for maintaining the correct concentration of potassium and sodium in the fluid that surrounds body cells. The hormones that achieve this are called mineralocorticoids.

Reproduction

The timing of reproductive development in all vertebrates and most invertebrates is under hormonal control. Similar hormones that control sexual maturation occur in many different animal groups. For example, annelid worms have gonadotropin hormones very similar to those found in humans.

All animals have some endocrine control of reproduction. Worms of many groups have tissues in their brains that secrete hormones to control the development of gonads (sex organs). Nereidine is a hormone that controls both growth and the timing of sexual development in annelids. Other types of worms have hormones that control the timing of egg maturation.

Neurohemal gonadotropic hormones are responsible for the timing and development of sexual characteristics and sexual maturity in a variety of invertebrates, including mollusks. Glands in the eyestalks of crustaceans produce gonadotropins. Crustaceans' androgenic glands make peptide hormones that lead to the development of male sex organs; degeneration of this gland leads to the development of female sex organs. In insects, an endocrine gland in the brain produces a hormone that stimulates the manufacture of vitellogenin, a protein used by the insect's developing oocytes (eggs) to make egg proteins.

Reproduction in vertebrates

Lancelets are marine animals that are not vertebrates but look very like their ancient ancestors. Lancelets have glands similar to the vertebrate pituitary gland. The glands produce the steroid luteinizing hormone (LH), which stimulates the production of sperm in male lancelets. LH is important in vertebrates, too. It is made and released by the pituitary gland. LH is a gonadotropin; it stimulates the development of the gonads, or sex organs, and leads to the production of the sex hormones testosterone, in males, and estrogen, in females.

Other vertebrate gonadotropins include androgens and progesterone. In mammals, for example, progesterone is essential for helping embryonic young survive in the uterus.

▼ **ESTROGEN CYCLE**
Mammal
The release of eggs in mammals is called ovulation. The pituitary gland secretes follicle-stimulating hormone. This stimulates the ovaries to release a hormone called estrogen, which controls the development of an egg follicle. The pituitary gland also releases LH, which causes an egg follicle to break open, releasing the egg into the fallopian tube. There, it may be fertilized.

GENETICS

Fungal hormones

All multicellular organisms have reproductive hormones. Even fungi have them. Many fungi go through a complex "alternation of generations" during their life cycle, with an asexual (budding) stage followed by a sexual stage that involves fertilization by sperm. During the asexual stage, fungi are haploid: they contain only one set of genes. To move into the sexual stage, two haploid cells must unite to form a diploid cell—a cell with a full, double set of genes, as found in the body cells of most other animals, including humans. Fungi accomplish this change using hormones. They produce peptide hormones that ensure attraction and fusion of different haploid cells.

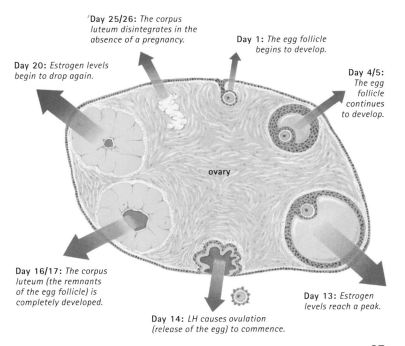

Day 25/26: *The corpus luteum disintegrates in the absence of a pregnancy.*

Day 1: *The egg follicle begins to develop.*

Day 20: *Estrogen levels begin to drop again.*

Day 4/5: *The egg follicle continues to develop.*

ovary

Day 16/17: *The corpus luteum (the remnants of the egg follicle) is completely developed.*

Day 13: *Estrogen levels reach a peak.*

Day 14: *LH causes ovulation (release of the egg) to commence.*

Vertebrate endocrine systems

Human

*The human endocrine
system is very similar to
that of other mammals.
It also has much in
common with the
endocrine systems of
non-mammalian
vertebrates.*

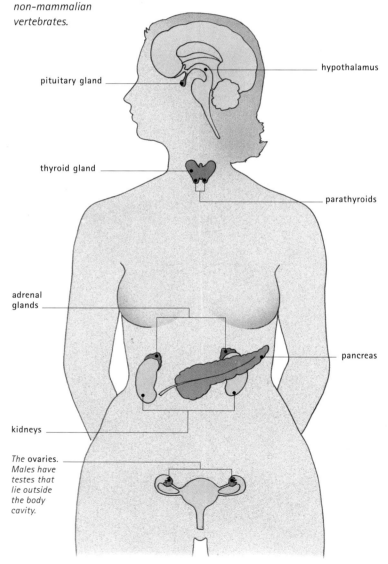

pituitary gland

thyroid gland

adrenal
glands

kidneys

*The ovaries.
Males have
testes that
lie outside
the body
cavity.*

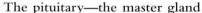

hypothalamus

parathyroids

pancreas

The basic structure of the endocrine system is similar in most vertebrates. In humans and most other vertebrates, the hypothalamus is located in the center of the base of the brain. The hypothalamus is a communication gateway, linking the nervous and endocrine systems. It contains cells that secrete chemicals called neurohormones. Some neurohormones trigger further hormone release, but others are inhibitory and prevent the secretion of other hormones. The hypothalamus receives signals from the body, then uses its neuro-hormones to signal to the next important gland—the pituitary gland.

The pituitary—the master gland

When it receives a neurosecretory signal from the hypothalamus, the pituitary gland releases one of its hormones. This may act on tissues directly or, more usually, may activate the production and release of other hormones from other glands. The pituitary is made up of two parts. The anterior (front) pituitary is like other glands, and secretes peptide hormones. The posterior (hind) pituitary is different. An extension of the hypothalamus, it consists mostly of axons (long extensions) of nerve cells originating in that part of the brain. All neurohormones from the hypothalamus go directly to the anterior pituitary and control all of its hormonal secretions.

Though it is controlled by the hypothalamus, the pituitary gland is often called the "master" gland. The pituitary synthesizes and releases a wide range of hormones, which influence all cells and most life processes in the body. The growth hormone, somatotropin, which targets liver and fat cells, originates in the pituitary. Somatotropin is crucial in stimulating the production of some other hormones needed for growth that are made by the liver and in other tissues. As growth requires energy, somatotropin is key in the metabolism of proteins, fats, and carbohydrates such as sugars. The timing and control of growth hormone release are regulated by a peptide hormone released from the hypothalamus, GHRH. GHRH also controls the release of a hormone that inhibits growth, called somatostatin, which is released from the hypothalamus, the pancreas, and several other endocrine organs.

Under orders from the hypothalamus, the anterior pituitary also produces thyroid-stimulating hormone, or TSH. TSH targets the thyroid gland. It causes proteins in the thyroid gland to be changed into the hormones T3 and T4. These hormones are involved in regulating the body's metabolic rate, growth and development, calcium levels, and the onset of puberty in humans.

Other pituitary hormones

Luteinizing hormone (LH) and follicle-stimulating hormone (FSH) are gonadotropins that stimulate the development of the sex organs and their hormones. LH and FSH are released by the anterior pituitary and are essential for reproduction in vertebrates. Prolactin is another pituitary hormone. In mammals, prolactin triggers mammary gland development and stimulates milk production during pregnancy and after birth.

Oxytocin is a peptide hormone formed in the hypothalamus. It passes to the posterior pituitary gland before moving into the bloodstream. Oxytocin stimulates the release of milk from the mammary glands of lactating female mammals. It also stimulates uterine contractions during childbirth, and may enhance bonding between a mother and her newborn young. In males, oxytocin enhances sperm mobility to increase the chances of fertilization. Although only mammals produce milk, many other vertebrates secrete oxytocin-like peptides. They stimulate the muscular contractions needed for laying eggs.

Another hormone produced in the pituitary is ACTH (adrenocorticotropic hormone), which acts on the cortex of the adrenal glands. The hypothalamus initiates the production of

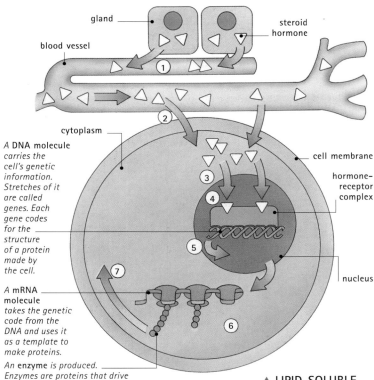

A DNA molecule carries the cell's genetic information. Stretches of it are called genes. Each gene codes for the structure of a protein made by the cell.

A mRNA molecule takes the genetic code from the DNA and uses it as a template to make proteins.

An enzyme is produced. Enzymes are proteins that drive chemical reactions in the cell.

ACTH in response to stress. ACTH stimulates the adrenal glands to produce glucocorticoid hormones, such as cortisol.

The adrenal glands

In mammals, each adrenal gland has two parts. The inner medulla produces hormones such as catecholamines, epinephrine, and

▲ LIPID-SOLUBLE HORMONES

Steroid hormones, such as glucocorticoids and mineralocorticoids, are lipid-soluble. This characteristic allows them to diffuse through cell walls to bond with receptors within cells. A steroid hormone is secreted by a gland into the bloodstream (1) to a target cell. It then diffuses across the cell membrane (2), passing through the cytoplasm (3) to bind with a receptor at the nucleus (4). The hormone-receptor complex causes certain parts of the cell's DNA to be expressed (5). This leads to the production of enzymes (6) that alter cellular activity (7).

What controls endocrine activity?

The effect of a hormone depends largely on its concentration in the bloodstream at any one time. The amount of hormone encountered by a cell is influenced by three factors. The first is the rate of production of the hormone. This is controlled by feedback loops. As hormone levels begin to rise, mechanisms are triggered that cause them to fall. If there is too little of a hormone in the bloodstream, more will be produced.

The second factor affecting a cell's exposure to a hormone is its rate of delivery—a high blood flow delivers more hormone molecules than a low blood flow. The third factor is the hormone's "half-life." This is the rate of breakdown and elimination of the hormone. If a hormone is degraded rapidly and has a short half-life, its levels will drop rapidly. However, some hormones have long half-lives and can have effects long after secretion has stopped.

norepinephrine. It is surrounded by the cortex, which secretes two types of hormones: glucocorticoids and mineralocorticoids.

Catecholamine production is stimulated by stress: danger, fear, anger, hypoglycemia (low levels of blood sugar), exercise, and trauma. Epinephrine is another adrenal medullary hormone that "kicks in" under stressful conditions. The release of catecholamines has a dramatic effect on the body, increasing the heart, metabolic, and breathing rates, and stimulating the burning of fat to yield extra energy. Catecholamines also inhibit functions that are not essential at times of stress, such as digestion. Catecholamines break down rapidly after they are released.

Glucocorticoids and mineralocorticoids, the hormones produced by the adrenal cortex, are steroids; like all steroids, they are made from cholesterol. Glucocorticoids are crucial for regulating the body's metabolism of glucose, a kind of sugar. Every body cell contains receptors for glucocorticoids. Glucocorticoids also have strong anti-inflammation properties (they help reduce swellings). Cortisol, also known as hydrocortisone, is a well-known glucocorticoid that is a commonly used anti-inflammatory drug. Mineralocorticoids are hormones that are vital for regulating the concentrations of potassium and sodium in the fluid that surrounds body cells.

▼ BLOOD SUGAR
Vertebrate
The level of glucose, a sugar, in the blood is regulated by two hormones produced by the pancreas: insulin and glucagon. When the amount of glucose in the blood gets too high, insulin acts to reduce the level. Glucagon does the reverse, causing the level of glucose to increase when it gets too low.

IN FOCUS

Digestive hormones

Digestion is a complex process, closely regulated by the nervous and endocrine systems. Some of the gastrointestinal (GI) hormones that control digestion are produced in the brain, but many others are made and secreted by the digestive system itself. Important GI hormones include gastrin, which is secreted by the stomach and regulates gastric acid secretions; cholecystokinin, which is produced by the small intestine and stimulates the production of enzymes and bile in the pancreas; and secretin, another small intestine hormone that leads to the production of digestive fluids in the pancreas and liver.

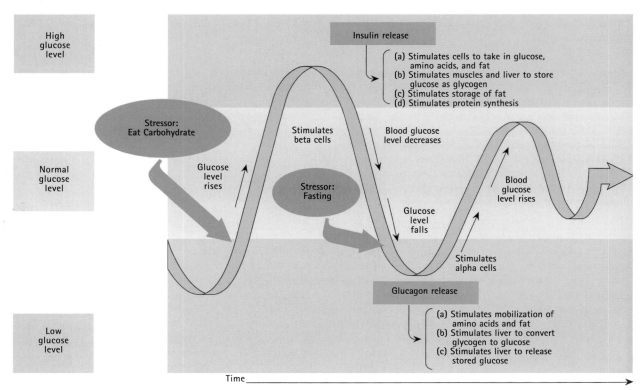

High glucose level

Normal glucose level

Low glucose level

Stressor: Eat Carbohydrate

Glucose level rises

Stimulates beta cells

Stressor: Fasting

Blood glucose level decreases

Glucose level falls

Insulin release
(a) Stimulates cells to take in glucose, amino acids, and fat
(b) Stimulates muscles and liver to store glucose as glycogen
(c) Stimulates storage of fat
(d) Stimulates protein synthesis

Blood glucose level rises

Stimulates alpha cells

Glucagon release
(a) Stimulates mobilization of amino acids and fat
(b) Stimulates liver to convert glycogen to glucose
(c) Stimulates liver to release stored glucose

Time

Hormones of the pancreas

The human pancreas is unusual in that it has both endocrine and exocrine functions. The pancreas is located next to the entrance to the small intestine. The endocrine part of the pancreas contains small clusters of cells called the islets of Langerhans, which produce several hormones. One of the hormones—somatostatin—inhibits growth; the other two are glucagon and insulin.

Insulin is a peptide hormone vital for the metabolism of carbohydrates and fats, and involved in the breakdown of proteins and minerals. Glucose is produced by the digestion of carbohydrates in the small intestine. When glucose moves into the bloodstream, insulin is released. Insulin enables cells in muscle and other tissues to take in glucose and use it as a source of energy. Insulin is also required to make the liver store glucose in the form of glycogen, so insulin lowers glucose concentrations in blood by allowing its uptake by cells around the body. As blood glucose levels fall, insulin production stops. Cells then cannot take up glucose, so they switch to alternative sources of energy, such as fatty acids. When blood sugar levels increase, insulin is again produced.

Glucagon works in the opposite way to insulin. This peptide hormone increases blood glucose levels by making the liver break down its glycogen reserves. Glucagon secretion is stopped by high levels of glucose in the blood.

Insulin and glucagon have similar functions in other vertebrates. All tetrapods (four-legged vertebrates) have discrete pancreatic glands. Fish, however, have pancreatic cells distributed widely through their digestive tract.

The pineal gland

The pineal gland, or epiphysis, is a small gland located in the brain. Shaped like a pinecone, the pineal gland secretes the hormone melatonin, which communicates information about light conditions in the environment to the body. Light that enters the eye is transmitted as a signal to the hypothalamus. This communicates the information via the spinal cord to the pineal gland.

The nerve signal triggers the production of melatonin from another chemical, serotonin. Melatonin production is lowest during the

The third eye

In amphibians and in many species of reptiles, the pineal gland is photosensitive; it is directly sensitive to light, although it cannot form images. It is often called the parietal eye, or third eye. The parietal eye is particularly well developed in lizardlike reptiles called tuataras, which live only on islands off New Zealand. They have a hole in the skull called the parietal foramen that allows light to reach the gland.

Photosensitivity allows the pineal gland to regulate biological rhythms by nervous signals to other regions of the brain. Direct photosensitivity has been lost in mammals, but the pineal gland remains important in regulating these rhythms. It is supplied with information on light levels by nervous signals originating in the visual cortex of the brain.

day when light levels are at their maximum. Melatonin output reaches a peak at night. The hormone helps regulate the body's circadian (daily) rhythms, including sleep–wake cycles and levels of daily activity. It also affects reproduction by inhibiting the production of the gonadotropins that lead to the development of sex organs and hormones. Irregularities in melatonin output cause insomnia and are also linked to seasonal affective disorder (wintertime depression).

▼ *Milk production is triggered by prolactin, which is released by the pituitary gland. Suckling causes the release of further hormones that maintain milk output.*

Exocrine system

simple tubular

simple coiled tubular

simple branched tubular

simple branched tubular (with ducts)

simple acinar

simple branched acinar

simple branched acinar

▶ EXOCRINE
GLANDS
The range of exocrine glands that occur in mammals.

CONNECTIONS

COMPARE the pheromone glands of an *ANT* with those of a *LION.*

COMPARE the poison production mechanisms of a cane toad with those of a *TARANTULA* and of a *SCORPION.*

Unlike endocrine glands, most of which release their hormones directly into the blood, the glands of the exocrine system release the chemicals they produce onto the surface of the body or into cavities inside it, such as the gut. Most exocrine glands secrete their products through ducts. They are generally simple tubular structures through which secretions flow, though some exocrine glands contain branching ducts of various sizes. Exocrine glands are classified by the type of secretion they release. Serous glands have cells that produce and release serous, or clear, watery secretions, such as tears. Mucus glands produce heavier mucus secretions. Mixed glands, such as the salivary glands, contain both kinds of cells, and their secretions are a combination of both.

Salivary glands

The salivary glands secrete chemicals that begin the digestion of food. In humans, the numerous salivary glands are scattered about the mouth and on the lips, on the insides of the cheeks, and on the tongue. However, there are three major pairs of these glands. The parotids are the largest; they are located in front of and below the ears. The sublingual glands are on the floor of the mouth under the tongue. The submaxillary glands are located under the jaw. The salivary glands release saliva, a mixture of water, ions, mucus, and digestive enzymes.

Saliva has many important functions. It aids in the digestion of food, since it contains amylase, an enzyme that breaks down starches

into sugars; and it helps bind food into a slippery bundle that is easily swallowed. Saliva also helps keep the body cool. Dogs and some other animals pant when they are hot; the evaporation of saliva cools blood passing through vessels close to the tongue's surface.

The sweat glands

Maintaining a constant temperature is essential for warm-blooded animals, so most also lose heat through the evaporation of sweat. Perspiration is released from sweat glands, traveling through ducts and out of pores to the surface of the skin. In humans, overstimulated

IN FOCUS

Territory markers

Vertebrates use pheromones as sexual attractants and for territorial marking. The females of many mammals produce pheromones during estrus to attract mates. Some male mammals use pheromone-laced urine to mark their territory. For example, the dik-dik (a type of African antelope) has scent glands at the inner corners of its eyes. By touching a gland to a twig or branch, the dik-dik marks the boundaries of its territory. Male sungazer lizards also use pheromones; theirs are released through pores on the underside of the thighs.

nerves also trigger the release of sweat. Sweat glands are connected to nerves of the sympathetic nervous system, so they respond to emotional states, such as fear. Fear-induced sweating is stimulated by the release of epinephrine from the adrenal glands.

The lachrymal glands

Humans and most other vertebrates have lachrymal glands—better known as tear ducts. They produce serous secretions that maintain moisture on the eyes. The lachrymal glands are almond-shaped and are located at the outer corner of each eye. A conjunctival mucus gland adds a little mucus to tears.

Human eyes are constantly washed by a film of tears, which keeps them lubricated and clean. The film has three layers—an oily layer, a watery layer, and a mucus layer. All are essential for the health of the eye. Blinking helps maintain an even coating of tears over the whole of the eyes exposed to air.

Attack and defense

Animals use exocrine secretions for a range of purposes: for attracting mates, for establishing territories, in defense, or for hunting. Many insects, including female moths, produce and release chemicals called pheromones to attract mates. Neurohormones released by an insect's brain cause specialized body tissues to release pheromones, which are also important for regulating group behavior among social insects, such as ants.

Exocrine secretions may be used for hunting prey. Some snakes, for example, use venom to immobilize prey. Cobras have venom glands all along their jaw and fangs. The venom is delivered through the hollow needlelike fangs. Many other animals use exocrine secretions for defense. The red lionfish has glands at the base of each of its sharp spines. When an attacker comes into contact with a spine, the gland releases a potentially lethal chemical. Bombardier beetles have a pair of glands near their anus that produce a spray containing a cocktail of hot, corrosive, and toxic chemicals. Skunks, too, are famous for their chemical defenses. The distasteful odor of a threatened skunk is produced by secretions from the animal's anal glands.

NATALIE GOLDSTEIN

Marvelous milk

One of the most important exocrine secretions is milk. Found only in mammals, milk is secreted by the female's mammary glands. It is full of nutrients that help a young mammal grow quickly. Milk varies greatly in composition between species. For example, species with young that need to lay down a thick layer of blubber as soon as possible after birth, such as seals, have milk that is extremely rich in fat. Seal milk contains up to 60 percent fat.

▼ Mammary glands

In humans, milk emerges from the nipple through several ducts. In ungulates (hoofed mammals), milk is discharged through a single duct.

Human

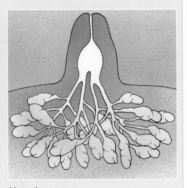

Ungulate

FURTHER READING AND RESEARCH

Silverthorn, Dee. 1998. *Human Physiology: An Integrated Approach.* Prentice Hall: Upper Saddle River, NJ.

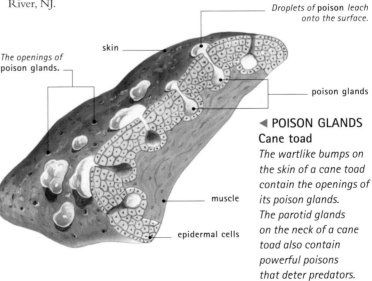

Droplets of poison *leach onto the surface.*

skin

The openings of poison glands.

poison glands

muscle

epidermal cells

◄ POISON GLANDS
Cane toad

The wartlike bumps on the skin of a cane toad contain the openings of its poison glands. The parotid glands on the neck of a cane toad also contain powerful poisons that deter predators.

Immune, defense, and lymphatic systems

From the instant an animal is born, and all through its life, its body is under threat from disease-causing organisms, or pathogens. Humans and other vertebrates come under attack; so do invertebrates. Pathogens are very diverse in size, in what they contain, and in what they have evolved to do. Since animals' bodies are not impenetrable, they are invaded almost continually. Yet most life-forms manage to stay healthy most of the time. This is because of their immune system, a complex system of physiological mechanisms that has evolved to identify, attack, and destroy most enemies. Most of the time, the immune system is successful.

The enemy

There are an enormous number of disease-causing organisms that can enter and infect the body of others. Among them are bacteria, single-celled organisms without a nucleus; protozoans, single-celled organisms with a nucleus; viruses, minute packets of DNA and protein, which do not have the other structures or equipment of a cell; various types of fungi; and larger organisms such as parasitic worms.

SYSTEM HIGHLIGHTS

INNATE IMMUNE SYSTEM Innate immunity is the system of immunity which animals are born with and which is not altered by their experience of fighting pathogens. This system includes the protective effects of the skin, phagocytic cells, and mucus on membranes. *See pages 77–79.*

ADAPTIVE IMMUNE SYSTEM The adaptive immune system evolves as a result of an animal's experience of pathogens. The two main groups of adaptive immune cells are B cells and T cells. The adaptive immune system produces antibodies that destroy invading pathogens. These antibodies evolve by a process of mutation, becoming increasingly effective at fighting off particular infections. *See pages 80–85.*

LYMPHATIC SYSTEM A system of ducts that permeates the body and recycles and cleans body fluids. This fluid contains immune cells that identify and destroy foreign antigens. *See pages 86–89.*

AUTOIMMUNITY AND GENETIC DISORDERS Sometimes the immune system malfunctions and attacks an organism's own cells. *See pages 90–91.*

Pathogenic organisms cause harm in a variety of ways. Some poison the body with the toxins they produce: for example, the bacteria that cause diphtheria in humans live in the trachea and produce a dangerous toxin that locally disorganizes the function of the tissues. Other pathogens, notably viruses, subvert the normal functioning of cells. Some parasites merely rob the body of some of the nutrition it needs. Viruses must get inside cells before they can reproduce and do harm. Some bacteria also infect animals' cells: for example, the mycobacteria that cause tuberculosis in humans. Others kinds of bacteria, however, live in the spaces between cells—either within tissues or in body cavities.

Often, as with tuberculosis (TB), it is the body's own reaction to the presence of certain pathogens that results in disease. The mycobacteria that cause TB are difficult to kill; and once they have entered cells, the host's defense systems begin damaging its own cells and tissues in order to deal with the infection.

Body cells are constantly bathed in a complex blend of chemicals carried by the blood. Cells must be selective about which of these chemicals they let in or respond to, and which they keep out or ignore. Every body cell has surface receptors that act as gatekeepers, controlling the responses of the cell. Pathogens that infect body cells do so by binding to receptors on the surface and inducing the cell to pull them inside.

Often, disease begins when pathogens successfully bind to the surface of body cells. Successful binding occurs more often on mucosal surfaces, such as the lining of the nose or digestive tract, than on the skin. In the case of viral infection, this first binding allows an individual virus particle to enter and subvert the normal function of a cell. Information carried by the virus may then induce the host cell to manufacture more identical virus particles. However, some infections follow a more subtle and complex course.

Inflammation

Inflammation is a frequent symptom of many infectious diseases. In the case of human tuberculosis, mentioned above, a lot of the damage caused by the disease results because the human immune system attacks cells infected with the mycobacteria. The attack creates chronic (long-lasting) inflammation of the respiratory tract.

HOW INFECTION OCCURS

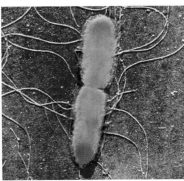

The bacteria that cause pneumonia (such as Streptococcus pneumoniae) are transmitted by small droplets of water vapor in the air. They cause inflammation of the throat and lungs.

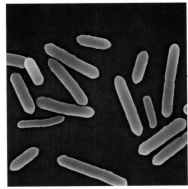

The typhoid bacterium (Salmonella typhosa) enters the intestines through infected food or water and causes a serious generalized illness that is fatal in 25 percent of all cases.

The bacterium that causes tetanus, or lockjaw, enters the body through any wound. Tetanus is characterized by muscle rigidity and spasms.

▲ SITES OF INFECTION

The site at which an infection enters the body depends upon the type of invading organism. Infections may result from bacteria, protozoans, viruses, fungi, or worms.

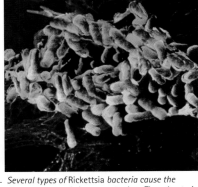

Several types of Rickettsia bacteria cause the disease commonly known as typhus. These bacteria are carried by ticks, fleas, lice, and mites.

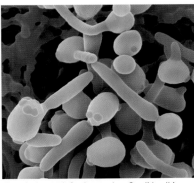

A genital yeast infection, such as Candida albicans, is transmitted by sexual contact or by infected towels and in women induces a vaginal discharge.

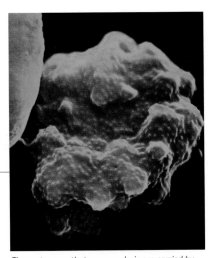

The protozoans that cause malaria are carried by mosquitoes. The protozoans are transmitted to humans by mosquito bites. This red blood cell has been infected by the malaria protozoan Plasmodium falciparum.

Some acute (short-lasting, but violent) inflammation results when the immune system deals successfully with the local attack of a pathogen. For example, when staphylococci bacteria enter a hair follicle in an animal's skin and begin to multiply, the animal will develop some soreness, reddening, and swelling as the inflammatory process locally increases blood supply. Lymph (a pale fluid containing the white blood cells of the immune system) also flows into the local tissue, which then swells. The inflammation is useful for the animal under attack because it identifies the enemy, increases the local forces to deal with the enemy, and keeps the defense forces well supplied. Sometimes, a little white "head" of pus may then accumulate as the next stage of inflammation. Pus is a concentration of white blood cells known as neutrophils. If the neutrophils successfully deal with the staphylococci—as they usually do—all is well and the inflammation ceases as the threat is defeated.

Through the long course of evolution, animals have developed different strategies to counter the many kinds of assault by pathogens. The pressure of natural selection on pathogens living in hostile environments (other organisms) that are trying to kill them has led to the evolution of a number of counterstrategies to avoid, evade, or disrupt animal immune responses. This kind of balance, where two kinds of organisms develop strategy and counterstrategy for exploitation or defense, is often described as an "evolutionary arms race." Similar arms races develop between predators and prey, parasites and hosts, and many other kinds of competing life-forms.

Types of immunity

Immune mechanisms can be thought of as falling into two main types: innate immunity and adaptive immunity. The idea behind this division is simple. Innate immunity comprises the parts of the immune system that exist whatever the host's experience of disease organisms has been in its lifetime. In contrast, adaptive immunity (also called acquired immunity) is influenced by the host's experience of pathogens (or of immunizations given to protect against pathogens).

Adaptive immunity is specific and "learned." Having experienced a certain type of pathogen once, the immune system of each host "remembers" it and reacts harder and faster to the same type of pathogen on a later occasion. This is why many diseases (such as chicken pox) are once-in-a-lifetime occurrences, and why vaccination can be such an effective defense.

The variable features of a host's adaptive immune responses derive from the activities of a group of cells known collectively as lymphocytes. Unlike any other body cells, lymphocytes are able to construct new working genes from component parts of the host's genome (the genome is the total genetic information contained in chromosomes, and is functionally divided into genes coding for proteins). By effectively creating new genes, the lymphocytes can produce potentially unique versions of proteins—although these are constrained by a basic general design. The products of these genes are highly specialized proteins known as antibodies and T-cell receptors. Both antibodies and T-cell receptors are involved in the recognition of foreign material (matter that originates outside the host's body, such as a pathogen or debris that has entered the body as a result of injury).

Adaptive immunity is thought to be restricted to vertebrates, but many more animals appear to have some level of innate immunity. Innate immunity does not involve lymphocytes. It is a combination of generally protective activities performed by a range of other cells and tissues. All complex organisms, including insects, annelids, mollusks, and many much simpler life-forms, possess innate immune responses.

The recognition of what is foreign to a host's body, and thus potentially a pathogen, is vital for immune responses to work. Just as important is knowing what belongs to the host's body, in other words what is "self," and thus not normally a threat. The immune response is not perfect: the innate part falls short of recognizing all pathogens, and the adaptive part sometimes starts to damage the host. This second phenomenon is called autoimmune disease, and rheumatoid arthritis is probably the best-known example.

IN FOCUS

Insect immunity

Insects possess innate, nonspecific immunity. They have no lymphocytes and they do not produce immunoglobulins. However, they do produce pathogen-destroying chemicals, for example, proteins called cecropins. Cecropins can successfully combat some, but not all, bacteria that infect insects by rupturing or "lysing" the bacterial cell membranes. A second class of insect immune chemicals, the attacins, act against a limited number of bacteria that tend to infect insects' alimentary canal. Many insect species also possess a type of lysozyme, an enzyme that lyses bacterial cells. All known insect immune responses are nonspecific and involve chemicals that are more or less successful at defeating insect pathogens.

Innate immune system

Innate immunity begins with an organism's surfaces: in mammals, these are mainly the skin and the mucus membranes. The outermost layer of human skin, for example, has considerable mechanical strength, is composed of dead cells, and is slightly acidic. This all helps make it an effective barrier. However, it is not impenetrable: hair follicles, sebaceous ducts, and sweat glands can all offer points of entry for a pathogen. The importance of skin as a first line of defense becomes very obvious when it is damaged. When large areas of skin areas are lost after serious burns, for example, the underlying tissue is highly vulnerable to infection and needs expert dressing to protect it. Similarly, a penetrating wound is often a source of bacterial infection.

Compared with skin, the moist, warm, mucous membranes are an attractive target for pathogens. Membranes' basic defense is the fluid coating of large sugar-rich mucin molecules on their surfaces. These molecules are kept moving by the beating of cilia in many areas (for example, in the trachea) or by peristalsis (in the intestines). The flow of mucus physically impedes the attachment of pathogens. Mucus also inhibits pathogen attachment chemically. Many viruses are able to bind onto complex sugars on the cell surface. Similar sugar residues (for example, N-acetyl galactosamine) in the mucus compete to bind the virus and bear it harmlessly away before it reaches the cell surface.

Interferon

In human tissues, one aspect of innate immunity is shared by almost all cells. Cells attacked by viruses produce a warning protein called interferon, which binds to receptors on nearby cells. As its name suggests, interferon acts as a signal to "interfere" with, or inhibit, viral replication in these cells. If a cell that has received an interferon signal subsequently becomes infected, it is already primed to "go slow" with regard to protein synthesis (viruses need protein synthesis to happen fast) and to

make an enzyme to destroy ribonucleic acid (RNA), a chemical vital in gene expression. Without RNA, the virus cannot be replicated. Signaling molecules similar to interferon may also cause infected cells to "commit suicide."

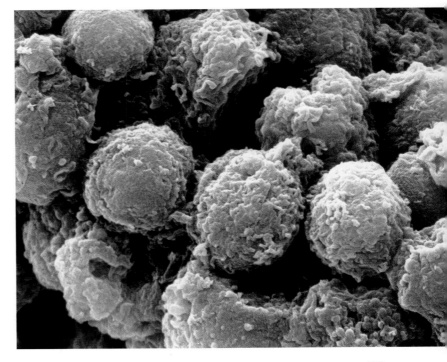

▼ These monocytes are white blood cells that have formed in the bone marrow. They are phagocytic, engulfing and digesting invading pathogens.

All systems go!

The action of antibodies is normally considered under the heading of adaptive immunity, but antibodies also have functions in the innate immune system. Scientists tend to divide the immune system into innate and adaptive mechanisms for purposes of understanding, but in reality when it comes to dealing with infection, everything works together. Antibody molecules can refine and focus not just the binding of molecules in the complement cascade but also the activities of phagocytic cells and of those leukocytes such as eosinophils that kill without engulfing. So, despite their origin in the adaptive immune system, antibodies do not generally achieve a separate solution to infectious disease. Instead, they are part of an integrated assault on pathogens from both innate and adaptive systems.

their recognizing the pathogen involved. A host that does not know an enemy has entered cannot take steps to remove it.

Animals have receptors on cell surfaces and on free molecules in body fluids to immediately recognize various molecular patterns or shapes that are characteristic of pathogen surfaces. Bacteria, fungi, and viruses all have these distinctive surface patterns, and all complex animals have receptors that can recognize at least some of them. For example, bacteria typically have a coat containing a chemical called lipopolysaccharide (LPS). LPS possesses a molecular pattern that sets alarm bells ringing in the innate immune systems of everything from humans to beetles or snails. When a host's defensive cells bind LPS on their surfaces, they are activated to start a sequence of defensive measures.

Toll-like receptors

Other important molecules whose patterns are recognized by the innate immune systems of most animals include the protein flagellin (so called because it is found in the whiplike flagella that bacteria use for movement) and various bacterial forms of nucleic acids. Like LPS, all these pattern molecules are recognized

Cell suicide is called apoptosis. This emergency measure can be a useful weapon against some pathogens: dead cells are useless to invading viruses and thus tend to stop the infection from spreading.

Inevitably, infections will sometimes start to take hold. A huge range of innate defenses can begin to operate within an organism that is under threat, but a great deal depends upon

▶ **INNATE IMMUNITY**

In the natural immune system, invading pathogens such as bacteria are ingested and destroyed by cells called neutrophils and macrophages.

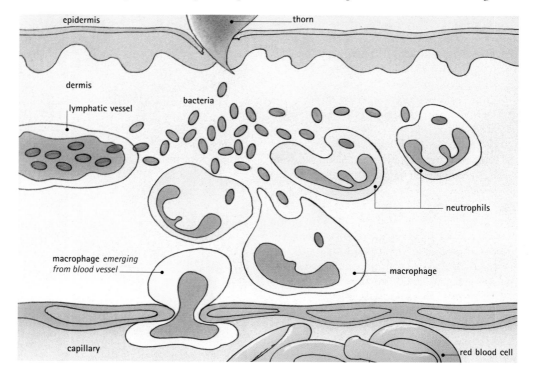

by a family of receptors called toll-like receptors, or TLRs, which are found on animals' innate immune cells. In mammals, TLRs are well represented on types of cells called dendritic cells, which are embedded throughout the animal's tissues, especially near body surfaces. Although the role of dendritic cells is not fully understood, it seems that when the TLRs recognize a warning sign in the form of a particular molecular pattern, the dendritic cell sends out a signal in the form of a type of protein called a cytokine. Cytokine release coincides with the start of both inflammation and immunity. Dendritic cells act as sentry guards, continuously sampling their local environment and poised to trigger local defense mechanisms as soon as a pathogen is detected in the vicinity.

The complement cascade

Dendritic cells may not provide the only warning signals. Not all pattern-recognizing receptors are attached to cell surfaces. Some such as the mannose binding lectin (MBL), which recognizes the bacterial surface sugar mannose, are soluble and exist free in the host's body fluids. MBL sticks to mannose and begins an immune reaction called a "complement cascade." The reaction leads to the formation of small molecules called anaphylatoxins, which rapidly diffuse away from the site and act in several ways. As with the cytokines produced by dendritic cells, anaphylatoxin release can be associated with an accumulation of neutrophils (pus cells), and changes to the local blood supply, bringing more defenses to the infected area. The anaphylatoxins also induce capillaries to leak plasma into the affected tissue, making it swell.

The complement cascade (so called because it supports, or complements, the action of antibodies) is an immune reaction involving about 30 proteins that interact and modify one another. Initially these interactions take the form of a positive feedback loop: that is, the initial response is repeatedly magnified and can rapidly lead to a major reaction. In addition to the release of anaphylatoxins, the cascade also leads to the formation of a "membrane attack complex," which helps destroy pathogens by punching holes in their cell walls. It also triggers the deposition of a product called C3b

on the surface of pathogens. This makes the particle much more likely to be taken up, and therefore destroyed, by a phagocytic cell. This process is called opsonization.

Leukocytes

In addition to neutrophils, the cells of the innate immune system also include several other types of white blood cells, or leukocytes, which may work separately or together to kill an invader. The protective roles of mast cells, basophils, and eosinophils are probably connected with killing large pathogens, such as parasitic helminth worms. Neutrophils are phagocytic leukocytes (named for the Greek *phagein*, meaning "to eat"). They patrol the body tissues, on the lookout for pathogens to engulf and consume. Macrophages ("big eaters") are cells that routinely collect and consume cell debris but also engulf any pathogens they may find. Neutrophils and macrophages are capable of killing most bacteria, which they engulf in a membrane-bound organelle called a phagosome. In the cell's cytoplasm, organelles called lysosomes fuse with the phagosome, producing a phagolysosome. The lysosomes deposit powerful chemicals into the space containing the bacteria. The bacteria are attacked by these compounds, which include free radicals and destructive enzymes. Some pathogens can withstand this kind of treatment, but it is generally effective at killing bacteria.

Leukocytes also contain peptides (molecules that are structurally similar to proteins but smaller) called defensins, which bind into and damage the cell walls of bacteria. Defensins are certainly important, but their role in the innate immune system is not fully understood.

CONNECTIONS

COMPARE the immune system of a *CRAB* with that of a *HUMAN*. Crabs are probably just as able as a human to fight off pathogen attacks, but they defend themselves differently. Human blood contains about 30 different proteins that can release anaphylatoxins to help destroy disease-causing pathogens. Horseshoe crabs rely on just two proteins, limulin and alpha 2-macroglobulin, to accomplish this task.

IN FOCUS

Crybaby

Tears are an innate immune mechanism that protects the surface of the mammalian eye by creating a flow to impede pathogen binding. Tears also contain a good antibacterial agent called lysozyme. This acts in a way similar to some antibiotic medicines, killing bacteria by attacking their cell walls.

Adaptive immune system

Antibodies (sometimes known collectively by the singular term "antibody") are the immune system molecules that most people have heard of. Antibody is produced by one of the great classes of lymphocytes (called B lymphocytes, or "B cells"). B cells not only secrete antibody but also bear antibody molecules on their surface membranes. These surface immunoglobulins, or sIgs, are representative of the specific antibody the B cell would secrete into the blood serum (the liquid component of blood, in which blood cells are suspended) if the cell was turned on. Antibody molecules are proteins with some sugar residues attached.

The most common kind of antibody, known as immunoglobulin G (IgG), has a basic shape similar to the letter Y. The base or foot of the Y (known as the Fc region) is where an antibody can dock to and interact with a host cell, or interact with proteins of the complement system. The tips of the arms of the Y (the molecule's "hands") are where the antibody binds with its target or "antigen," usually a foreign material. The precise structure of the "hands" varies greatly, thus allowing different antibody molecules to bind specifically to different antigens, or to different parts of the same antigen. The scientific name for the molecule's "hands" is complementarity determining regions (CDRs). Roughly, their

CLOSE-UP

Types of antibodies

The basic Y antibody is called IgG (Ig is short for immunoglobulin). There are various other, less common antibody types. For example, IgM looks like five Y's arranged in a star with their bases joined together, and IgA can resemble a single Y or two Y's joined tail to tail. Antibody molecules are found in the blood serum and lymph, and so bathes the tissue spaces. Huge amounts of IgA are made and poured onto mucous surfaces such as those of the intestines.

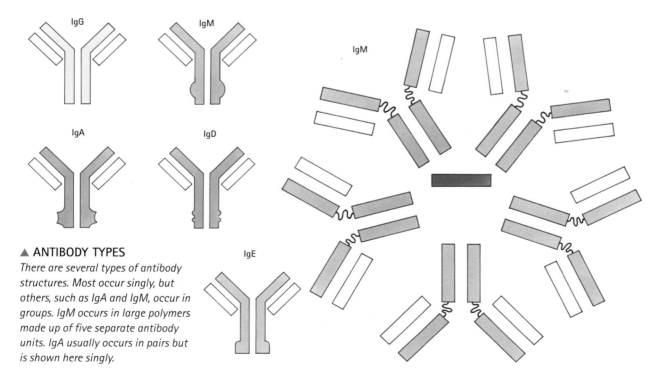

▲ ANTIBODY TYPES
There are several types of antibody structures. Most occur singly, but others, such as IgA and IgM, occur in groups. IgM occurs in large polymers made up of five separate antibody units. IgA usually occurs in pairs but is shown here singly.

molecular shapes are "complementary" to the molecular shape of the antigen concerned, so they fit together in the same precise way as a lock and key. In fact, think of those keys where you can see a flat surface with hollows in it; inside the lock is a flat surface but with bumps that complement the hollows. Antibody-antigen binding works in a comparable way.

The function of antibodies

The simplest role of antibody molecules is neutralization or blocking. A microbe, especially a virus, is neutralized if antibody binds to the molecules it would otherwise use to latch onto and invade a cell. The antibody acts as a simple blocker, preventing the virus from linking up with a cell membrane. Antibody molecules can also neutralize toxins. Many pathogenic bacteria produce chemicals that are poisonous to host cells. An antibody molecule that binds to a biologically essential part of the toxin can render it nontoxic.

Another important antibody activity is agglutination. Agglutination is the process by which an antibody molecule links two other molecules or cells, forming a bridge between them. They are in effect stuck together with the antibody as glue. IgM, with its fivefold starlike array of Y-shapes, can cross-link as many as 10 antigen sites. Pathogens or chemicals held together like this may be disabled.

A third crucial role of antibody molecules is opsonization. Any antigen that has been bound by an antibody molecule, regardless of whether or not it has also been agglutinated or neutralized, is marked out for destruction by phagocytic cells. In this respect antibody is working somewhat like C3b. The opsonizing activity works because many phagocytic cells have receptors on their surfaces for the foot, or Fc region, of the antibody molecule. These receptors are thus known as Fc receptors; this is where the antibodies "dock."

So in general, antibodies operate separately from the lymphocyte cell that produces them, attacking pathogens in the body fluids, or docking on the surface of a different cell type, such as a phagocyte, and influencing its activity.

Phagocytes dispose of pathogens by engulfing them, but other types of cell have different means of attack. Some cells kill pathogens by shooting out toxic contents onto a target such

▼ ALLERGIC RESPONSE
Some people are allergic to antigens called allergens. Allergens provoke a response in the immune system that may, for example, cause an unnecessary release of histamine, a chemical normally associated with the immune response to infection. This results in symptoms such as swelling, irritated nose and throat, and watery eyes. A common allergic response is hay fever, in which the immune system responds inappropriately to the presence of pollen.

allergens enter body

Allergens attach themselves to a **white blood cell**

Allergens stimulate white blood cell to change into **plasma cell**

A plasma cell makes **antibodies**

allergens enter body for second time

Allergens and antibodies combine, and **histamine** *is released*

Antibodies attach themselves to a **mast cell**

as a pathogen or an infected cell. Usually the toxins act by rupturing or "lysing" the target cell. In this instance the antibody molecules direct the attack by binding to the target and linking it to the killer call. For example, if one of your own cells was infected by a virus that was replicating, then killing the infected cell might be the best way to prevent many neighboring cells from becoming infected. This carefully targeted poisoning of marked cells is called antibody-dependent cell-mediated cytotoxicity (ADCC). There are several sets of killer cells that work in this way. It is a highly effective system, but there can be unfortunate complications—a form of allergy.

A further very important role of antibody is that it can focus innate immune responses such as those of the complement cascade system, just as it focuses ADCC. When an antibody

ADCC

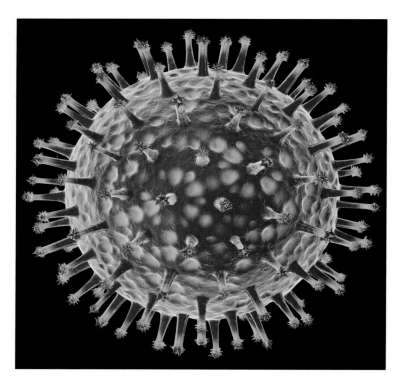

molecule binds to its antigen, a subtle change occurs in the shape of the foot or Fc region. One of the complement proteins is attracted to this changed shape, and the interaction triggers the cascade system. The complement cascade acts hard and fast. Using its positive feedback loop, it brings all the vigor of the innate response to bear directly on the culprit organism, summoning an army of neutrophils and initiating other physiological responses. In effect, the innate immune system is a blunt weapon that can be honed to deadly accuracy by the precise ability of the adaptive system to identify trouble. It is a beautiful example of cooperation between systems.

◀ *Viruses, such as this bird flu virus, can reproduce only inside a cell. They gain entry to the cell by binding to surface receptors and inducing the cell to pull them inside. Once inside, they alter the cell's processes, causing it to produce more virus particles. The cell, however, also produces a protein called interferon, which inhibits the further replication of the virus in surrounding cells.*

▼ **Antibody structure** *An antibody is composed of heavy and light chains of molecules held together by disulfide bridges. Each antibody unit has two antigen binding sites. These are the regions which are most structurally diverse and which evolve by processes of mutation and selection. Thus the adaptive immune system is able to respond to new pathogens and, in most cases, defeat them.*

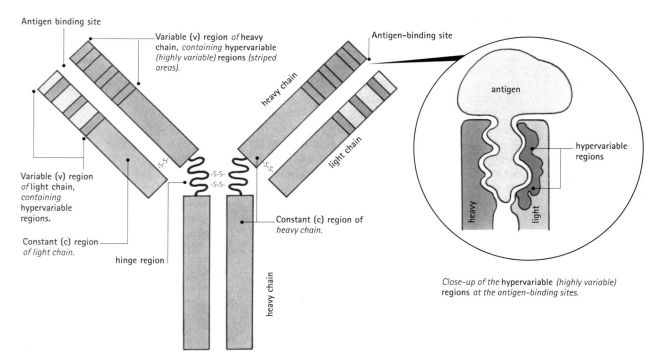

Antigen binding site

Variable (v) region *of* heavy chain, *containing* hypervariable *(highly variable)* regions *(striped areas).*

Antigen-binding site

heavy chain

light chain

antigen

hypervariable regions

heavy

light

Variable (v) region *of* light chain, *containing* hypervariable regions.

Constant (c) region *of light chain.*

hinge region

-S-S-
-S-S-

heavy chain

Constant (c) region of *heavy chain.*

heavy chain

Close-up of the **hypervariable** *(highly variable)* **regions** *at the antigen-binding sites.*

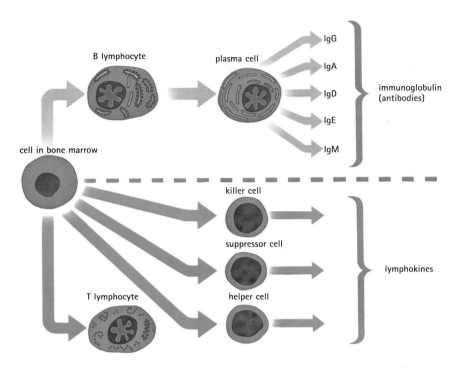

COMPARE the immune system of a vertebrate such as a **HUMAN** with that of an invertebrate such as a **STARFISH**. In addition to the innate immune system, a human has an adaptive immune system based on antibodies; but starfish, spiders, insects, and many other invertebrates have an immune system that is not based on antibodies.

CONNECTIONS

▲ IMMUNE CELL DIVERSITY

A bone marrow stem cell may divide into a B lymphocyte (B cell) or a T lymphocyte (T cell). These two types of immune cell differentiate further into specialized immune cells with specific functions in fighting disease.

Allergies

Mast cells are part of the immune response that probably evolved in response to large parasitic worms, whose cuticles (external surfaces) are very resistant to attack. Mast cells and eosinophils attack using the antibody-dependent cell-mediated cytotoxicity (ADCC) described earlier. Mast cells bear a form of antibody called IgEs. In parts of the world where parasitic worm infestation is routine, this manifestation of the immune system is fundamentally useful. Unfortunately IgEs are also frequently made in response to a set of antigens that are not directly pathogenic. We call these antigens allergens. Pollen, house dust mites, and cat dander (minute scales from skin or hair) are examples of common allergens. When an allergen cross-links an IgE molecule on a mast cell surface to another IgE molecule on the mast cell, the cell releases a cocktail of inflammatory molecules into the location, including the protein histamine, and an allergic episode is started. Because of the nature of the allergens, the site of first contact is likely to be an exposed body surface or airway—hence the itching, runny eyes and nose, and irritated throat experienced by allergy sufferers.

CLOSE-UP

Surface immunoglobulins

B-cell surface immunoglobulin (sIg) is a special case. This is a type of antibody found on the surface of B lymphocyte cells. It represents the specificity of antibody that B cells produce if they develop all the way into secreting plasma cells. sIg is crucial to the fate of the B cell. If sIg reacts with "self," the B cell will be programmed to die by an interaction with the cells that surround it in the marrow before it goes out into the circulation. If the cell passes this test (by not recognizing anything) the next job of sIg will be to stimulate the cell to divide, if and when antigen that matches the sIg is encountered, during an infection.

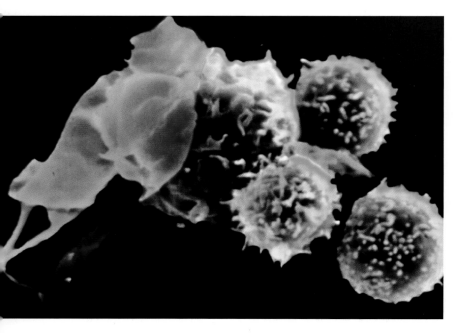

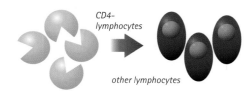

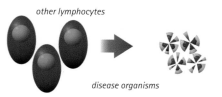

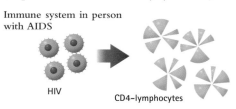

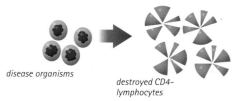

HIV and AIDS

When the human immunodeficiency virus (HIV) enters target cells, it injects its genetic material into the cell nucleus, where the viral genes are spliced directly into the cell's own DNA. Thereafter, every time the cell replicates its own genome and divides, it also copies HIV.

For 7 to 10 years after infection, the body produces helper T cells, which mobilize cytotoxic T cells to fight the infection. No one knows why, but infected people have no symptoms during these early years. The immune system battles ferociously, but in most cases, without the assistance of medication it ultimately loses.

At some point, HIV mutates in such a way that infected dendritic cells pass HIV on to helper T cells in the lymph nodes. HIV kills the helper T cells. The body boosts its production of helper T cells in efforts to activate B cells and killer T cells, but the more helper T cells are produced, the worse the infection becomes and the more helper T cells are killed by the virus. When the level of helper T cells in the blood falls below a certain level, the infected person's immune system is very seriously impaired, and the person has AIDS (acquired immune deficiency syndrome).

▲ *A colored scanning electron micrograph (SEM) of four killer T lymphocytes (shaded blue) attacking a cancer cell (shaded red). The killer T lymphocytes recognize the cancer cell by its surface antigens.*

Allergies can have other origins apart from IgE and mast cells, but this form, with its rapid onset, is referred to as immediate hypersensitivity. Its most severe manifestation is anaphylactic shock, which can be quickly fatal. The allergen in this reaction is likely to have been injected: for example, an insect venom.

▶ **HIV**
HIV destroys the CD4-lymphocytes, which help to control the action of other lymphocytes. Thus the immune system is impaired and may not be able to rid the body of some disease-causing organisms, leading to a combination of illnesses diagnosed as AIDS.

Normal immune system

disease organisms   **CD4-lymphocytes**

CD4-lymphocytes are alerted when disease organisms enter the body.

CD4-lymphocytes

other lymphocytes

CD4-lymphocytes help control the response of other lymphocytes.

other lymphocytes

disease organisms

The other lymphocytes destroy the disease organisms.

Immune system in person with AIDS

HIV   **CD4-lymphocytes**

HIV multiplies inside the CD4-lymphocytes and may destroy them.

disease organisms   *destroyed CD4-lymphocytes*

The body's immune responses may fail when disease organisms invade because the CD4-lymphocytes have been destroyed.

disease organisms

The disease organisms may overwhelm the immune system.

Major histocompatability complex

One apparent drawback of a highly developed adaptive immune system is the difficulty it poses for tissue and organ transplantation. For example, if an organ like a kidney is randomly taken from one individual and implanted into another individual it will have little chance of surviving, because it will be recognized as foreign by the recipient's immune system and rejected. Through the study of transplantation success and failure (for example, with skin grafts between mice), scientists were able to understand tissue compatibility, which is also called histocompatibility.

Among the many different molecules present on the surface of cells are proteins of the major histocompatibility complex (MHC). These proteins act as advertisements, showing what is happening inside the cell. They can signal "business as usual," or indicate that something different is happening within. There are two different types of advertising molecules: MHC class I and MHC class II molecules. MHC class I molecules, which are displayed on the surface of most cells, contain peptides that represent the kinds of compounds the cell is making. MHC class II molecules, on the other hand, display peptides that originate from what the cell is degrading. Not all cells bear MHC class II molecules—generally these advertisements appear on specialized cells associated with immune function, such as those in the lymph nodes. Usually, both the manufactured peptides advertised by MHC class I molecules and those being broken down and advertised by MHC class II molecules will be bits of "self" protein. Their manufacture and disassembly are part of normal cell activity. Thus when MHC I and MHC II molecules both display "self" peptides, the signal is "business as usual," and the cells attract no attention.

However, if a tissue cell is infected by a virus, its protein synthesis mechanisms are partly taken over, and peptides that are definitely "nonself" will be produced and displayed on the surface by MHC class I molecules. This attracts the attention of cytotoxic T cells, whose receptors (TCRs) are "tuned in" to changes in MHC class I structure. On detecting foreign peptides, the T cells turn into potential killers and the cell making virus is likely to be destroyed. The "tuning in" of cytotoxic T cells to MHC class I depends on another cell surface molecule called CD8. MHC class I molecules can alert cytotoxic T cells to the presence of antigens in almost any cell in an animal's body.

If a cell bearing MHC class II molecule signals the destruction of a foreign material within, the immune response is rather more complicated. MHC class II molecules are recognized not by cytotoxic T cells but by helper T cells. These have the molecule CD4 on their surfaces. Helper T cells will not kill the cell carrying the foreign MHC class II advertisement, but having been alerted, they start to divide and produce molecules that will stimulate neighboring B cells into producing more antibody. B cells are likely to have begun gathering in the area already, having separately recognized the foreign antigen. Within a lymph node, any foreign-derived material arriving may be detected in two ways: directly by B cells and indirectly as pieces of peptide by helper T cells.

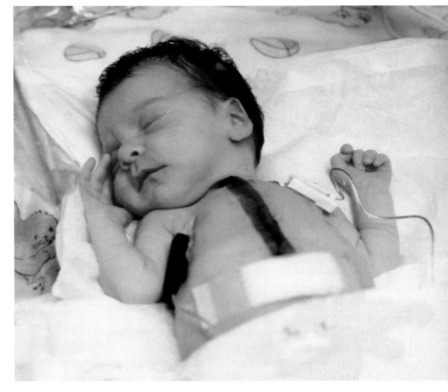

▼ This baby was born with a serious heart defect and required a heart transplant. To reduce the risk that a transplant organ will be rejected by the immune system, the organ must be "typed" to ensure that its antigens are as similar as possible to those of the recipient. In addition, drugs that suppress the immune system must be administered.

Lymphatic system

The lymphatic system consists of a series of ducts, or lymphatics, connected to numerous small masses of tissue called lymph nodes. The function of the lymphatic system is to collect, sample, and recycle body fluid. As blood circulates, its plasma—clear, colorless liquid containing oxygen, proteins, glucose, and other nutrients—seeps through capillary walls to refresh the extracellular fluid between cells. Over time, plasma can either seep back into the bloodstream or be drained by the lymphatic system. The lymphatic system carries this fluid (now called lymph) through the lymphatics until it reaches one of the lymph nodes. Lymph nodes filter material from the lymph and test it for certain constituents. Filtered material includes the degraded remains of the host's own dead cells, but the lymph may also include foreign matter draining from the site of an infection. Lymphocytes (T cells and B cells) gather in different parts of the lymph nodes, and when receptors on specific lymphocytes bind to the foreign antigens, an immune response to the infection is launched. Almost all of this activity depends on major histocompatability complex molecules. The activity in a lymph node during infection can be enough to make the node become enlarged and tender. For example, a person may have swollen "glands" (lymph nodes) in his or her neck when suffering a streptococcal throat infection. Lymphatic vessels are located in most parts of the body, but slightly different structures serve in the drainage of mucosal surfaces such as the gut and bronchi.

Origin of lymphocytes

Lymphocytes are "born" in bone marrow, the soft tissue that fills many of our long bones. Bone marrow contains stem cells that are constantly dividing to produce immature lymphocytes, which are released into the bloodstream. Immature lymphocytes do not carry specific receptors but develop such receptors when they mature. B lymphocytes mature in a range of sites including the bone

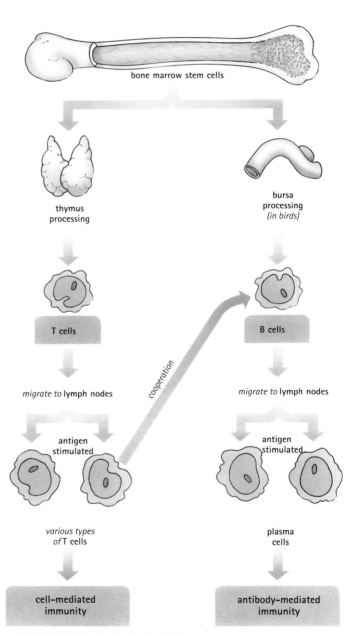

▲ PRODUCTION OF LYMPHOCYTES

B cells and T cells originate in the bone marrow, where they are produced by stem cells. B cells mature in the bone marrow, except in birds, where they mature in a lymphatic organ called the bursa of Fabricius. T cells pass to the thymus, where they mature into helper T cells and cytotoxic T cells.

LYMPHATIC SYSTEM

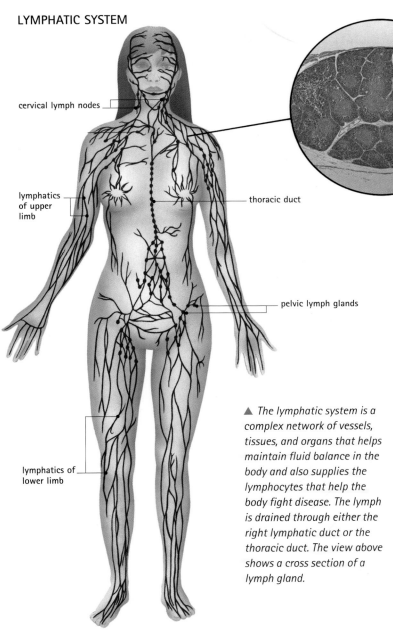

T cells and cytotoxic T cells and acquire receptors that enable them to detect specific antigens. Any thymocytes having receptors that could react with "self" material are eliminated. So rigorous is the testing of thymocytes that only about 5 to 10 percent of all thymocytes created become full-fledged T lymphocytes and move on to life in the immune system.

Generation of diversity

There is a fundamental similarity in the way that the huge numbers of diverse new receptor molecules are made in both B and T lymphocytes. This diversity is achieved through creating new working genes by recombining segments from several fairly limited sets of component gene pieces—a bit like a jigsaw puzzle in which the pieces can be put together different ways to make different pictures.

Why do we need an adaptive immune system at all? We have innate immunity to a number of pathogens recognized by the toll-like receptors (TLRs) of the dendritic cells. In theory the number of different TLRs could be extended to include every sort of foreign molecule. In reality such a system would be colossally inefficient. Not only would the number of genes required to provide full innate protection be unthinkably huge, but the majority of the receptors would never be used—even the most unfortunate individual is likely to encounter only a fraction of the potential pathogens that exist in the world. Not only would such a system require an impossibly large genome; only one new, unrecognizable pathogen could defeat the whole system.

Consequently, the system that has evolved in vertebrates (with a primitive form in the jawless hagfish, and a more advanced form in sharks) is an adaptive system in which cells are able to make a large variety of receptors by recombining a relatively small number of gene segments—just as we can build an enormous number of words from a set of just 26 letters. These cells also induce mutation in limited, but functionally crucial, regions of the newly assembled genes. The range of potential

▲ *The lymphatic system is a complex network of vessels, tissues, and organs that helps maintain fluid balance in the body and also supplies the lymphocytes that help the body fight disease. The lymph is drained through either the right lymphatic duct or the thoracic duct. The view above shows a cross section of a lymph gland.*

Labels on diagram:
cervical lymph nodes
lymphatics of upper limb
thoracic duct
pelvic lymph glands
lymphatics of lower limb

marrow; T cells mature in the thymus gland—a part of the lymphatic system situated in the neck or chest. The human thymus gland is located in the chest, just behind the breastbone. Infants have a large thymus, but the gland dwindles after puberty. Within the thymus, T cells multiply repeatedly as they move through the network of thymus cells. T cells present in the thymus are called thymocytes. During this process, the thymocytes differentiate into helper

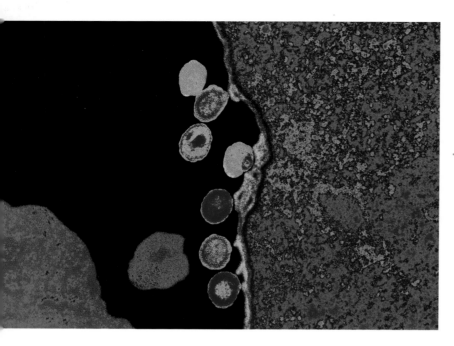

▲ *Virus particles of HIV leaving a cell. HIV may remain present in memory T cells, preventing drugs from eradicating the virus from the body.*

binding molecules is vast, but only those that are needed to deal with a clear threat are chosen, and the impact on the size of the genome is very modest.

Molecules of antibody, or immunoglobulin (Ig) are made in B cells. In an immature B cell one gene segment is selected at random for every two or three parts needed to build each of two peptide chains. The number of gene segments available for each part varies from 4 to about 65, so the total number of permutations in a given molecule is large, producing about 3 million potential antibody shapes. For a surface immunoglobulin (sIg), the completed protein will have two of each of the two peptide chains. This assembling of gene segments to arrive at a new compound gene is called somatic recombination. It is achieved by specific enzymes, and the sites at which they work are tightly controlled. The result will therefore always be recognizably a gene for an Ig.

Fine-tuning

Thus far, the mechanisms for producing Ig are very similar to those involved in TCR production. However, Ig molecules produced in B cells undergo a further two rounds of mutation that generate still greater specificity in their ability to bind antigen. The enzymes that control somatic recombination also introduce new nucleotides, that is, ones not found in the germ-line DNA (that which the animal inherits from its parents). These small alterations increase the diversity of compound genes produced by about 30 million times. Crucially, these nucleotide mutations affect sections of the gene that are eventually expressed as the CDRs, the highly variable "hands" of the Ig molecule, which consequently bind to specific antigen.

Thus the immune system has an overall repertoire of about 90 trillion (9×10^{13}) different antibody molecules. A young B cell will generate just one of these specificities, and will start by displaying it as sIg. Some of these young B cells will be deleted; others will never be needed and will therefore never give rise to any descendants.

Further changes in specificity will occur once an antibody is found to be useful. If a fit is made between sIg and an antigen experienced during infection, the B cell in question begins to divide, and its progeny will secrete more of the same antibody. The adaptability of the system does not end there.

Tolerance and clonal selection

Because the generation of antibody and TCR binding specificity is essentially random, the body will inevitably sometimes produce antibodies or TCRs that recognize its own cells or molecules. The adaptive immune system must "weed out" such agents before they have a chance to proliferate and begin a destructive attack on "self" tissues. In other words, the immune system must be continually kept in check to ensure that it remains tolerant of "self." Control is achieved by a process called clonal selection. Selection of T cells takes place in the thymus gland, which gets rid of any thymocyte whose receptor binds too well to "self" MHC, and also any which fails to bind to anything at all. Only cells that pass this test are allowed to go on maturing and dividing. In the bone marrow, a similar selection awaits B cells, based on their sIg's recognizing or not recognizing the "self" antigens in their environment. If the receptors on the sIg bind to "self," the B cell undergoes programmed cell death or apoptosis—effectively, it commits suicide.

In a process called somatic hypermutation, some activated B cells begin randomly introducing single nucleotide mutations into genes used to build the antibody in question. Many of these mutations will make the resultant antibody less of a match for the antigen. Cells with this hypermutation will fail to go on being stimulated by available antigen, and so production will cease. However some mutations may actually improve the fit the antibody can make with the antigen. B cells in which this happens will be maximally stimulated, and this new improved version will go into full production. The overall result of somatic hypermutation is that the antibody continues to be improved on and may become more effective as the immune response matures.

Cells with long memories

There is another crucial element in the response of lymphocytes: memory. At stages in the production and fine-tuning of an immune response, a subset of the successful cells will become quiescent and remain, not as functioning antibody secreting cells, but as "watchdog" cells ready to be stimulated

should the same antigen be encountered again in the future. As a result, the response to any further infections will be much better and faster than the first time, as the immune system has a head start in the form of mature B cells poised to produce the most effective antibody right away.

CLOSE-UP

Drifting and shifting

When the DNA in a virus mutates only slightly, the process is called antigenic drift. Large and dramatic mutations in the virus are known as antigenic shift. In the case of antigenic drift, the virus may still be genetically similar to the original. People may retain some antibody immunity against the virus, and a vaccine for the original form may provide some protection. When a virus undergoes antigenic shift, however, existing antibodies and vaccines are powerless to combat it because it is so different from its previous form. The influenza A virus frequently undergoes antigenic drift. Sometimes it also undergoes antigenic shift. Worldwide epidemics (pandemics) of influenza often follow an antigenic shift in the influenza A virus. In 1918–1919, at least 20 million people died in an influenza A pandemic. Influenza A pandemics occur every 10 to 40 years. The last one was in 1968.

▼ SIZE AND LOCATION OF THYMUS
The thymus is much larger in a child than in an adult, reflecting the importance of this gland in establishing the body's immune system early in life.

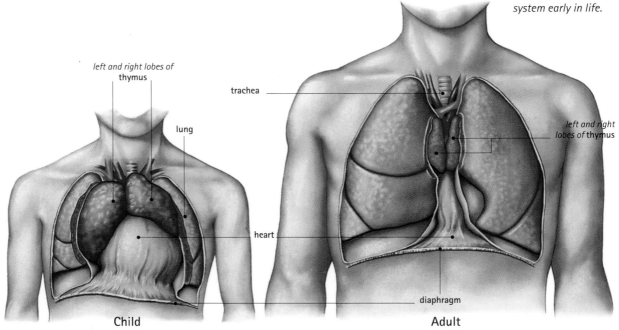

left and right lobes of thymus

lung

trachea

left and right lobes of thymus

heart

diaphragm

Child

Adult

Autoimmunity and genetic disorders

No biological system is perfect, and sometimes reactivity against self does occur. The unfortunate fact is that not all self-reactive B and T cells are excluded by clonal selection. The mechanisms that cause an immune system to attack the body it is supposed to defend are often far from clear, but a huge range of self material ranging from antibody molecules themselves to our own DNA can become targets of "autoantibodies." Similarly, T cells that damage self are frequently encountered. Some very serious and damaging diseases, including forms of diabetes and rheumatoid arthritis, are a result of autoimmunity. Surprisingly, the possession of autoantibody does not necessarily imply disease. Some autoimmune problems are temporary and seem to rectify themselves after a while. Thyroiditis, for example, which sometimes follows viral infection, is a short-lived disease, but the thyroid is also a frequent target of chronic (long-term) autoimmunity,

which can be highly debilitating. It seems likely that at least a proportion of autoantibodies are a result of similarity between the specificities generated by particular infections and some self antigen. For example, a common complication of Chagas' disease, or American trypanosomiasis, is autoimmune damage to the nervous system, and streptococcal rheumatic fever appears to stimulate autoantibodies against heart tissues.

A genetic link?

Some autoimmune conditions appear to be associated with particular variants of MHC molecules. Because MHC molecules are coded for by germ-line DNA, such diseases can be inherited. Not everyone with a particular MHC variant gets the disease, but the association can be very high—as in the joint disease, ankylosing spondylitis. The root of the disease may be the way particular peptide binding is handled by individuals' MHC.

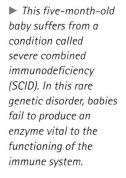

▶ *This five-month-old baby suffers from a condition called severe combined immunodeficiency (SCID). In this rare genetic disorder, babies fail to produce an enzyme vital to the functioning of the immune system.*

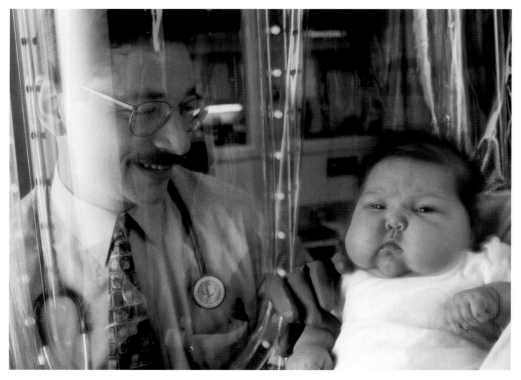

Unfortunately, some children are born lacking a component of the immune system. The first sign that an infant may have one of these rare congenital disorders will probably be the child's inability to fight a simple infection. Such disorders are called primary immunodeficiencies, and they include severe combined immunodeficiency (SCID), or "bubble boy disorder." Children with SCID have either no B cells and T cells or, if these cells are present, they have severe defects. Another congenital immune disorder is Wiskott-Aldrich syndrome (WAS). This syndrome is inherited through the X chromosome, so it affects only males, whose B and T cells can combat some infections but not others.

Apart from defects in lymphocytes, immunodeficiency can also be caused by numerous other failures: for example, the lack of a component of the complement cascade system, or a failure to produce the antibacterial toxins in killer granulocytes. People with genetic immune disorders may be treated with antibacterial and antiviral medications and are sometimes infused with purified antibody separated from donated blood. A number of attempts at gene therapy have been made: a modified virus is used to import a working copy of the defective gene into the patient's cells. This medical technology has proved unreliable, and sometimes dangerous, and it is now expected that stem cell transplantation will become the treatment of choice for many of these conditions.

GRAHAM MITCHELL/NATALIE GOLDSTEIN

FURTHER READING AND RESEARCH
Friedlander, Mark, and Terry M. Phillips. 1998. *The Immune System: Your Body's Disease-Fighting Army*. Lerner Publications: Minneapolis, MN.
Sompayrac, Lauren. 2003. *How the Immune System Works*. Blackwell: Malden, MA.
How your immune system works:
http://health.howstuffworks.com/immune-system.htm
Immune system:
www.niaid.nih.gov/final/immun/immun.htm
Immune system:
http://uhaweb.hartford.edu/bugl/immune.htm

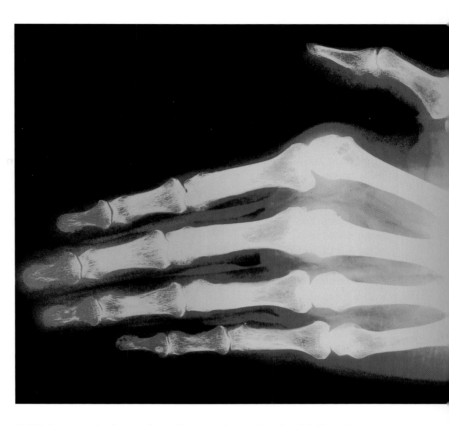

▲ *This heat-sensing image shows the warmth associated with inflamed, deformed, arthritic knuckles. Rheumatoid arthritis is an autoimmune disease in which the body's immune system attacks connective tissue, causing painful inflammation and stiffness in the joints.*

CLOSE-UP

One organ or more

There are two broad classes of autoimmune disorders: systemic and organ-specific. In systemic disorders, the immune system may attack multiple sites. The mechanism appears to involve a failure to clear immune complexes of antibody and self antigen from the circulation. One example of this is systemic lupus erythematosus, in which the immune system attacks joints, eyes, kidneys, and blood vessel walls. In organ-specific disorders, however, the immune system attacks only one specific type of organ. In the rare disorder called Addison's disease, for example, the immune system attacks only the adrenal cortex.

Muscular system

Muscles help animals move. At some stage in their life cycles, most animals are capable of moving their body from one place to another, perhaps to search for food, a mate, or a suitable place to settle. However, there are many other, more subtle movements that are equally important for the survival of an animal. These are the movements of individual body parts, such as the feeding appendages of barnacles, or internal organs, including the heart and the intestine.

Whatever the tissue or animal species, muscle cells are responsible for all movement in animals that involve more than just individual cells, and they account for almost every movement that can be observed with the unaided eye. Locomotion of individual cells, including unicellular eukaryotes, can be achieved by amoeboid movement or by beating of hairlike structures, such as cilia or flagella, but these will not be discussed here.

Contraction

Muscle cells are able to contract (shorten), and they can do this because they have many copies of two particular proteins: myosin and actin. All cells have versions of these proteins, but in muscle cells they are of particular types, especially abundant, and also organized into strands, or filaments. Myosin and actin filaments can slide against each other, and this is what happens throughout the muscle cell when it is contracting. There are many types of muscle cells; some work individually whereas others are organized into tissues and work together to form different types of muscles.

Muscles that carry out behaviors of the body are usually arranged in pairs or groups that work against each other. In limbs, for example, flexor muscles draw a part of the limb toward the body, and extensors straighten out the limb. To convert the force generated in a muscle to work,

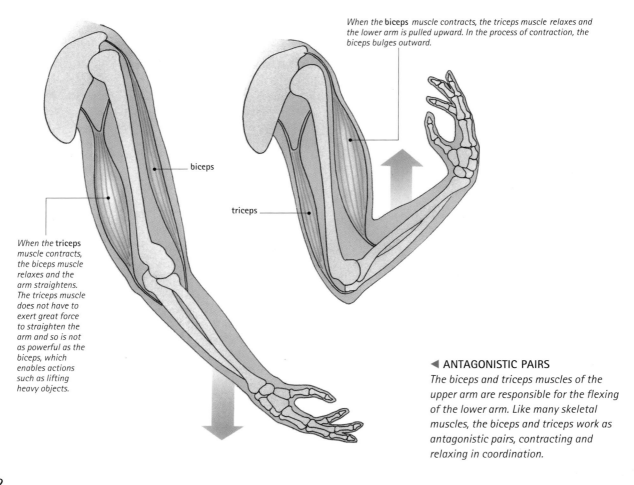

When the **biceps** muscle contracts, the triceps muscle relaxes and the lower arm is pulled upward. In the process of contraction, the biceps bulges outward.

biceps

triceps

When the **triceps** muscle contracts, the biceps muscle relaxes and the arm straightens. The triceps muscle does not have to exert great force to straighten the arm and so is not as powerful as the biceps, which enables actions such as lifting heavy objects.

◀ ANTAGONISTIC PAIRS
The biceps and triceps muscles of the upper arm are responsible for the flexing of the lower arm. Like many skeletal muscles, the biceps and triceps work as antagonistic pairs, contracting and relaxing in coordination.

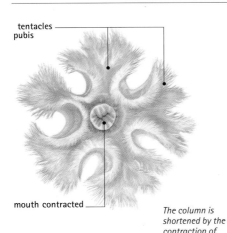

tentacles
pubis

mouth contracted

▲ CIRCULAR MUSCLES

The closing of a sea anemone's slitlike mouth is controlled by rings of circular muscles.

▶ SECTION THROUGH A SEA ANEMONE

A sea anemone does not have a hard skeleton. Instead, it has a hydrostatic skeleton: muscles contract against water held in compression by the anemone's tissues. The water is thus forced into other parts of its body, causing expansion or contraction.

The column is shortened by the contraction of longitudinal muscles.

The column is lengthened by the contraction of circular muscles.

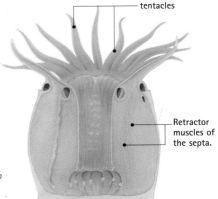

tentacles

Retractor muscles of the septa.

◀ EXTENSION AND CONTRACTION

The sea anemone uses longitudinal muscles that run perpendicular to the base and circular muscles parallel to the base to extend and contract the column, withdraw and extend the tentacles, and open and close the mouth.

the muscle must pull or press on a rigid structure of some kind. This structure could be an endoskeleton, such as our bones; or an exoskeleton, such as the cuticle of an insect or the shell of a clam.

Sometimes, the function of the muscles is to lengthen or shorten a structure that is not supported by a rigid skeleton. Muscles in sea anemones, for example, enable these soft-bodied animals to stretch out to catch food or retract to a small budlike shape. Layers of longitudinal and circular muscles work together to achieve such movements; contraction of the longitudinal muscles shortens the sea anemone, whereas contraction of the circular muscles

▲ *A network of tiny blood capillaries (red) supplies the muscles (brown) with the oxygen and nutrients they use to produce energy by the process of respiration.*

lengthens it. However, even in the case of the body movement of a sea anemone, the muscles are working against a "skeleton." This skeleton is the water that is retained in the hollow center of the animal. Because an enclosed volume of water does not compress if pressure is applied to it, the water inside a sea anemone functions as a hydrostatic skeleton against which the force of the muscles can be converted into work. In principle, this is also what happens in our heart muscle, but here the contraction of the heart forces the blood to pass into the circulatory system.

FEATURED SYSTEMS

HOW MUSCLES WORK Muscles are generally either smooth or striated. Voluntary muscles are striated, and involuntary muscles are smooth. *See pages 94–99.*

SKELETAL MUSCLE Used for functions such as locomotion, skeletal muscles are attached to skeletal structures such as the bones of vertebrates or the exoskeletons of some invertebrates. *See pages 100–103.*

CARDIAC MUSCLE This type of striated muscle has only one nucleus per muscle fiber. Contraction of the heart muscle is coordinated by the nerve bundles of the sinoatrial and atrioventricular nodes. *See pages 104–105.*

SMOOTH MUSCLE Smooth muscle lacks the obvious internal arrangement of striated muscles and are responsible for involuntary functions such as movement of food through the intestine. *See pages 106–107.*

How muscles work

COMPARE the arrangement of cells in smooth and striated muscles with the arrangement of cells in other tissues, such as the alimentary canal of the *DIGESTIVE AND EXCRETORY SYSTEMS*.

Muscles can be generally classified as either striated or smooth. In vertebrates, including humans, smooth muscles are responsible for many of the body functions we are normally not aware of, including regulating blood flow through blood vessels, expelling secretions from various glands, and moving food along the intestine. Striated muscles include both the heart muscle and the skeletal muscles; the latter move the bones of our skeleton in relation to each other. The term *striated* comes from the banded appearance of these muscles when viewed under a microscope. The bands are caused by the ordered organization of contractile myosin and actin filaments in striated muscle fibers.

Muscle structure

Structures within skeletal muscles were first studied using light microscopy and have therefore been given names according to their appearance. The primary repeated structural units of striated muscles are referred to as sarcomeres. Neighboring sarcomeres are divided by a prominent dense line of tissue called the Z disk. The so-called A bands span over the region of a sarcomere where the contractile myosin filaments are located. Each myosin filament is surrounded by thin actin filaments, which are anchored to the Z disk.

When a muscle is relaxed, the myosin filaments occupy the middle part of the sarcomere, leaving a region containing only actin filaments and the Z disk. This region forms a light band stretching from the end of the myosin filaments in one sarcomere to the beginning of the myosin filaments in the next. This lighter region is named the I band. The actin filaments are made up of globular G-actin molecules joined together like pearls on a string to form a double helix. In the vertebrate skeletal muscle, each myosin filament is surrounded by six actin filaments. The myosin

▼ **MUSCLE TYPES**

Muscles are divided into two main types: striated muscles, which make up skeletal muscles; and smooth muscles in, for example, the walls of the intestine, arteries, and veins.

▼ **CARDIAC MUSCLE**

In cardiac muscle, individual cells are branched and connected to one another by junctions called intercalated disks.

Striated muscle fibers

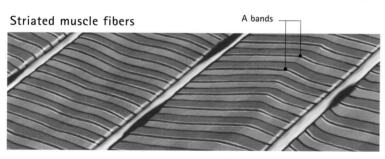

A bands

Smooth muscle fibers

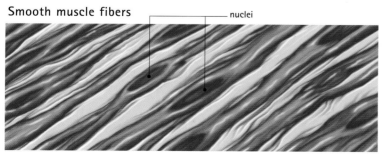

nuclei

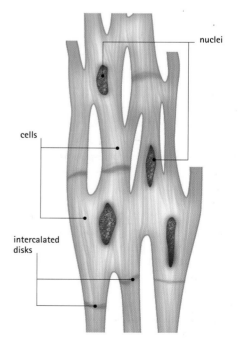

nuclei

cells

intercalated disks

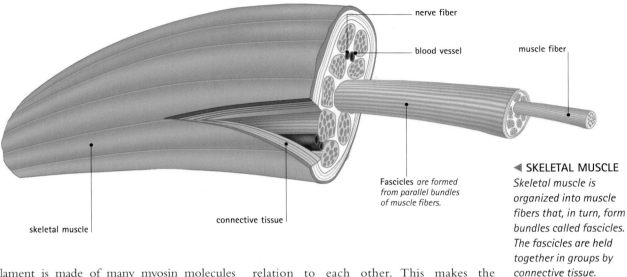

nerve fiber

blood vessel

muscle fiber

Fascicles *are formed from parallel bundles of muscle fibers.*

connective tissue

skeletal muscle

◀ SKELETAL MUSCLE
Skeletal muscle is organized into muscle fibers that, in turn, form bundles called fascicles. The fascicles are held together in groups by connective tissue.

filament is made of many myosin molecules bundled together. A myosin molecule contains a "head region," a "neck region," and a "tail region." It is made up of two long proteins and several small proteins located in the head region. The head of the myosin molecule contains the machinery that binds to actin and causes the contraction.

At the junction between the A and the I bands of the muscle fiber, there are numerous minute canals that tunnel from the contractile membrane—or sarcolemma—surrounding the muscle fiber to the interior of the muscle fiber. These canals are called transverse tubules (or T-tubules), which are continuous with the environment around the muscle fiber, make contact with every myofibril. The muscle fibers in striated muscles also have a well-developed sarcoplasmic reticulum, which is a baglike membranous structure corresponding to the endoplasmic reticulum of most other cells. The sarcoplasmic reticulum is located within the muscle fiber, and it stretches over each sarcomere. The sarcoplasmic reticulum is spread out over the myofibrils. Close to the T-tubules, it enlarges into terminal cisterns.

Cell biology of muscle contraction

It was first believed that the myosin and actin filaments became shorter during contraction of the muscle fiber, but that is not the case. Instead, the myosin and actin filaments slide in relation to each other. This makes the sarcomere shorter, but the length of the filaments themselves remains unchanged. What is happening is that the myosin heads are "walking" up the actin filaments, pulling them in line with one another. Two scientists, Hugh Huxley from the Massachusetts Institute of Technology and Andrew Huxley from Woods Hole Marine Biological Institute, introduced this sliding-filament theory independently in 1954. The theory has been revised many times, leading up to the current model.

CLOSE-UP

Muscle organization

Vertebrate skeletal muscles are composed of many bundles of long muscle fibers, which are the actual muscle cells. These bundles of muscle fibers are called fascicles, and within any fascicle all muscle fibers run in parallel. Each muscle fiber has many nuclei, because it forms through fusion of several immature muscle cells during development. Each muscle fiber contains numerous parallel elements called myofibrils. The myofibrils, in turn, consist of repeated units called sarcomeres. It is the sarcomere that is the functional contractile unit of the muscle.

FASCICLE

sarcolemma

myofibrils

The illustrations show progressive levels of detail from a fascicle to the molecular level. A fascicle contains many myofibrils, which are formed from alternating regions of thick and thin filaments that slide over one another, resulting in muscle contraction.

thick filament

thin filament

myofibril

T–tubules

mitochondria

sarcoplasmic reticulum

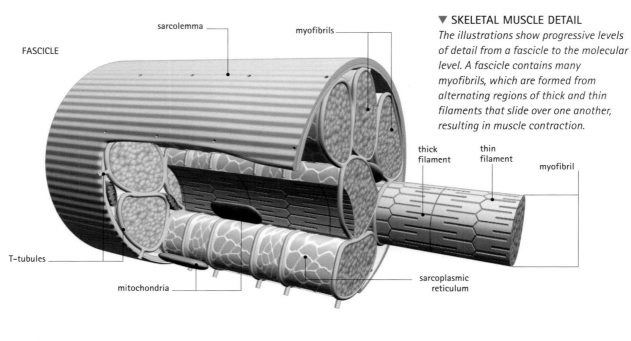

MYOFIBRIL

A band

sarcomere

myofibril

I band

H zone

Z disk

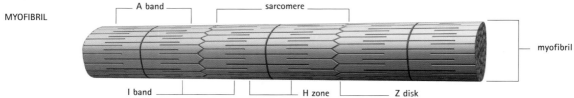

myosin heads

M line

thick filament

actin chain

Myosin heads "walk" along actin chains, contracting muscles.

thin filament

A band

Z disk

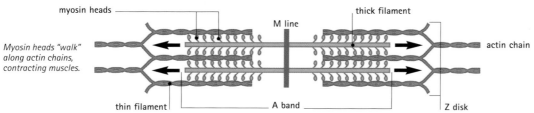

myosin head

M line

thick filaments

thin filament formed of actin chain

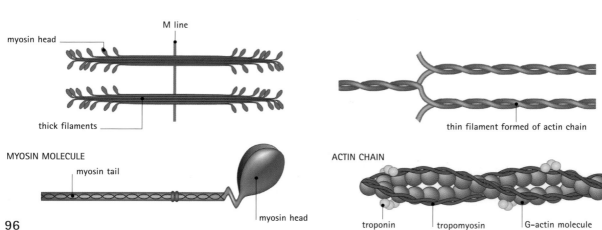

MYOSIN MOLECULE

myosin tail

myosin head

ACTIN CHAIN

troponin

tropomyosin

G-actin molecule

The sliding of actin and myosin filaments relative to each other leads to shortening of the sarcomere and contraction of the muscle. This process requires energy, and the energy is provided in chemical form by adenosine triphosphate (ATP), which is an energy-storage molecule found in all living cells. The ATP binds to the myosin head so that when the contraction cycle starts, the myosin head already has one molecule of ATP bound. The myosin head then binds to the actin filament; and with the energy released from the breakdown of the ATP to adenosine diphosphate (ADP), the myosin head turns from a 90-degree angle relative to actin to a 45-degree angle. Owing to this power stroke the myosin and actin filaments slide against each other. This is where the force of the contraction is created. During the power stroke, the ADP is released, leaving space for a new ATP to bind to the myosin head. Only once the ADP is replaced with a new ATP can the cross bridge between the myosin and actin filaments be broken and a new cycle be ready to start. In each cycle the myosin "walks" a little farther up the actin filament. The distance that the filaments slide against each other during the turning of one myosin head is tiny, but each sarcomere has many myosin heads that repeatedly go through the same cycle, and in a muscle fiber there are many sarcomeres stacked one after another. This makes the accumulated movement of myosin heads sufficient to shorten the muscle and move the body.

Role of calcium ions

In addition to energy in the form of ATP, filament sliding and muscle contraction need calcium ions (Ca^{2+}). An increased Ca^{2+} concentration in the cytosol (the semifluid substance that fills most of the space in a cell, except the organelles), or sarcoplasm, as it is called in muscle fibers, is the trigger that causes myosin heads to bind to actin filaments. Ca^{2+} starts the formation of myosin–actin cross bridges by binding to a protein associated with the actin filament. In the groove between the two intertwined actin strands, there is a long threadlike protein called tropomyosin. At distances of 400 nanometers (0.0004 mm) apart, representing a half turn of the actin

Rigor mortis

A little while after an animal dies, the body goes through a period during which it is stiff for several hours. This stiffness is called rigor mortis. After death, adenosine triphosphate, ATP, which is used for muscle contraction, is rapidly broken down throughout the body and disappears from the muscles. As ATP is needed for the myosin heads to release their bonds to the actin filaments, the muscles become fixed and rigid. The body remains in rigor mortis until the muscle decomposes to the extent that the myosin and actin filaments are destroyed.

double helix, there is a protein complex called troponin complex. The troponin complex is made of three subunits. One of these subunits binds actin, one binds tropomyosin, and one (the C-subunit) binds Ca^{2+}.

▼ To power the contraction of the muscles in these runners' legs, energy is released from large amounts of adenosine triphosphate, ATP. This release of energy enables myosin filaments to "walk" along actin chains in the muscle fibers, thus shortening the length of the muscles.

▼ CONTRACTION

Muscle contraction can be divided into six phases. (1) The myosin head is tightly bound to a myosin binding site at an angle of 45 degrees to the actin chain (labeled site a). (2) An ATP molecule binds with the nucleotide binding site, causing the myosin head to detach from the myosin binding site. (3) ATP is converted into adenosine diphosphate (ADP) and a phosphate. This action releases energy. (4) The myosin head rotates backward and attaches loosely to another actin site (labeled b) at an angle 90 degrees to the actin chain. (5) The inorganic phosphate is released, causing the myosin head to swing forward, thus pushing the actin chain forward. (6) The ADP is released, and the myosin head is once more bound tightly to the actin binding site.

When the muscle is relaxed, the tropomyosin covers the myosin-binding site on actin, preventing the myosin head from binding to it. If the Ca^{2+} concentration in the sarcoplasm is increased (to about 10 μm), four Ca^{2+} ions bind to the C-subunit of the troponin complex. Binding of Ca^{2+} to troponin C causes the whole troponin complex to retract, and in the process it pulls the tropomyosin molecule away from the myosin-binding site on each actin subunit. Now the myosin-binding site is exposed, and the myosin head can bind to the site, causing the sliding of filaments and, hence, contraction of the muscle to commence.

In a normal muscle fiber, the contraction will continue as long as sufficient ATP and Ca^{2+} are present close to the actin filaments. To end the contraction, Ca^{2+} must be removed from the sarcoplasm. To achieve this, the membrane of the sarcoplasmic reticulum is lined with Ca^{2+} pumps that move Ca^{2+} from the sarcoplasm back to the inside of the sarcoplasmic reticulum, which functions as a Ca^{2+} reservoir. Inside the sarcoplasmic reticulum Ca^{2+} binds to a special protein, which increases the reticulum's storage capacity of Ca^{2+}. The Ca^{2+} pumps are sufficient to keep the concentration of free Ca^{2+} in the sarcoplasm extremely low when the muscle is relaxed.

Action potentials

A contraction in a skeletal muscle starts with a series of electrical pulses (action potentials, or nerve impulses) traveling down a nerve to reach the individual muscle fibers. Each nerve cell, or nerve fiber, controls the contraction of several muscle fibers. Muscle fibers connected to the same nerve fiber contract and relax together; they are therefore called a motor unit. The nerve fiber ends in a nerve terminal that is very close to—but does not make direct contact with—the muscle fiber. This point of communication between nerve fiber and muscle fiber is a type of synapse called a neuromuscular junction.

In vertebrates, when the action potentials that have been traveling down the nerve fiber reach this terminal a chemical called acetylcholine is released from the nerve terminal. This neurotransmitter diffuses along the short distance from the nerve terminal to the sarcolemma of the muscle fiber. Embedded in the sarcolemma is a protein that binds the acetylcholine and functions as a receptor,

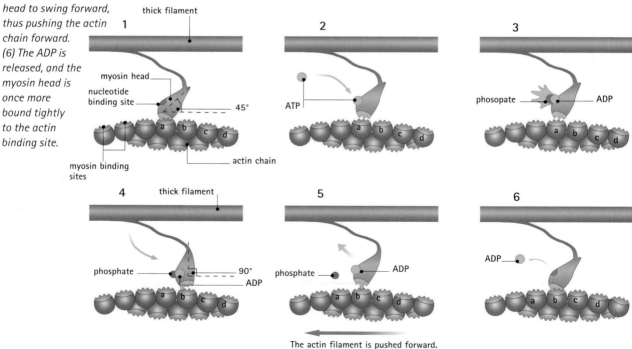

The actin filament is pushed forward.

mediating the signal to the muscle. Binding of acetylcholine to the acetylcholine receptor triggers the opening of an ion channel in this receptor protein. The channel allows positively charged sodium ions (Na^+) to enter into the muscle fiber. As a result of the inward flow of Na^+ ions, the inside of the muscle fiber becomes more positively charged. This charge triggers action potentials along the sarcolemma of the muscle fiber. When the action potential reaches a T-tubule, it follows the T-tubule into the interior of the muscle fiber.

In the membrane of the T-tubule, close to the terminal cisterns of the sarcoplasmic reticulum, there are voltage-sensitive proteins (dihydropyridine receptors) that change shape in response to the incoming action potentials. These voltage-sensitive proteins make direct contact with Ca^{2+} release channels (ryanodine receptors) located in the membrane of the sarcoplasmic reticulum. When stimulated by an action potential, the voltage-sensitive proteins in the T-tubule membrane force the Ca^{2+} release channels to open a pore through which Ca^{2+} is able to escape from the sarcoplasmic reticulum. As a consequence, the Ca^{2+} concentration of the sarcoplasm increases. This increase, in turn, results in binding of Ca^{2+} to troponin, cross-bridge formation between the myosin and actin filaments, active filament sliding, shortening of the sarcomere, and contraction of the muscle.

Energy for muscle work

Muscle cells need energy to carry out work; and, as noted above, this is provided in the form of energy-rich ATP. In animals ATP is produced primarily through the reaction between breakdown products of foodstuffs and molecular oxygen. Some muscle cells, such as those in the heart and in postural muscles, need an ample supply of oxygen to function. Other muscle cells can operate without oxygen for some time. During this time, these muscle cells make ATP from reactions that do not require oxygen. In this process, the body builds up an "oxygen debt," which later has to be repaid by a higher-than-normal oxygen uptake after the exercise.

Muscle cells that mostly rely on constant oxygen supply are typically reddish brown in appearance. This is for two reasons: first, they

contain many mitochondria, which are sausage-shaped organelles responsible for oxygen-dependent ATP production. Second, these muscle cells are typically rich in myoglobin, which is an oxygen-binding protein similar to hemoglobin. Myoglobin has a higher affinity for oxygen than does hemoglobin and is therefore able to "extract" oxygen from the blood and capture it for the muscle tissue. What we often refer to as red meat contains a high proportion of muscle fibers that require a rich oxygen supply; white meat has relatively more muscle fibers that can work without oxygen. In general, red muscle cells contract slowly but also fatigue slowly; white muscle cells contract quickly but also fatigue quickly.

▼ Nerve axons (the dark lines) connect to individual skeletal muscle fibers (the broad parallel bands) across a network of neuromuscular junctions, or motor end plates. The black dots are synapses. Magnified 400 times.

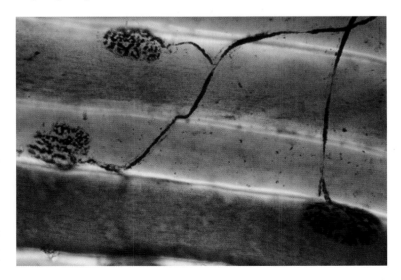

Skeletal muscle

COMPARE
the form and
arrangement of
skeletal muscles in
a fish, such as
a **TROUT,** with
their form and
arrangement in a
marine mammal,
such as a **DOLPHIN**
or **GRAY WHALE.**

CONNECTIONS

Although the basic mechanisms of most skeletal muscles are the same, differences in their structure and arrangement allow them to perform different jobs. Skeletal muscles in animals perform an astonishing range of tasks, including running, flying, sound production, and, in some fish, generation of electricity. Skeletal muscle attaches to bones in vertebrates and to the hard exoskeleton of many invertebrates. The locomotor muscles of an earthworm are adapted to create waves of contractions that spread along the body in a highly synchronized fashion. These contractions alternate between circular and longitudinal muscles to produce the waves of elongation and contraction that move the earthworm forward through the soil. The muscles in the claws of many decapod crustaceans, such as crabs, lobsters, and crayfish, are uniquely adapted to generate an incredibly strong force. These muscles are often short with a large cross-sectional area, and the fibers are arranged at an angle to the direction of the pull. Such an

arrangement produces a short range of motion but great power; it also allows the claw muscle to contract without bulging. This is an advantage because the space inside the claw is limited by the rigid exoskeleton.

Some muscles can contract and relax extremely quickly. For example, the muscles that operate the wings of some dipteran insects such as hoverflies can contract and relax several hundred times per second, providing the characteristic high-frequency wing beat of this animal group. Interestingly, this rate is much faster than that of action potentials in the nerve fibers that control the flight muscles. Researchers have found that once a cycle of alternate contractions in elevator and depressor wing muscles has been initiated, this cycle is propagated without the need for an action potential for every contraction. In these so-called asynchronous flight muscles the action potentials serve as an on and off switch, and the wings continue to beat so long as action potentials keep coming to the muscles.

▶ **INSECT WINGS**

An insect is an invertebrate, so it does not have an internal skeleton. Instead, an insect's hard exoskeleton serves as an anchor for muscles. The upstroke of an insect's wing is powered by the contraction of dorsoventral muscles. These muscles pull on the tergite (the top of the insect's thorax), causing the tergite to flex downward and the wings to flex upward. During the downstroke, the reverse happens. The dorsoventral muscles relax, and the basilar and dorsal longitudinal muscles contract, causing the tergite to be pushed upward and the wings to flex downward.

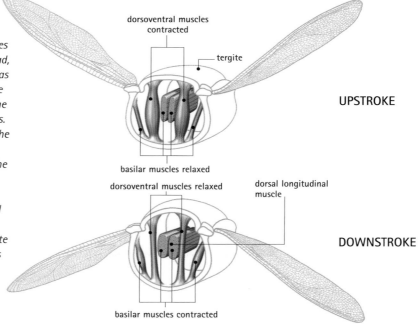

dorsoventral muscles
contracted

tergite

UPSTROKE

basilar muscles relaxed

dorsoventral muscles relaxed

dorsal longitudinal
muscle

DOWNSTROKE

basilar muscles contracted

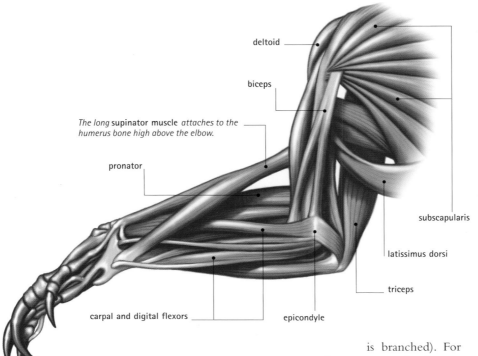

deltoid

biceps

The long **supinator muscle** *attaches to the humerus bone high above the elbow.*

pronator

carpal and digital flexors

epicondyle

triceps

latissimus dorsi

subscapularis

◀ ANTEATER ARM MUSCLES
The skeletal muscles of a giant anteater's forelimbs are particularly well developed. The anteater uses the strength these muscles provide to claw its way into rock-hard termite mounds.

As we have seen, muscles are adapted for different tasks. Some of these demand speed, others strength, and yet others endurance. One single type of muscle would clearly not be able to carry out such a variety of functions. Evolution has led to specialization of the skeletal muscles in different species as well as among the various muscles within an individual animal.

Arrangement of fascicles

Skeletal muscles are attached to the skeleton or other structures by tendons, and they generate movement by applying a force on the tendons, which in turn produces a pull on the attached bone. As noted previously, muscle fibers in a muscle are arranged in bundles, called fascicles. Within each fascicle, all muscle fibers are parallel to each other. However, as in the case of the crab claw muscle discussed above, fascicles are not always parallel. In fact, there are at least five different patterns of fascicle arrangements, each producing a specific property and usage of the muscle: parallel, fusiform, circular, triangular, unipennate (with muscle fibers on the same side of a tendon), bipennate (with muscle fibers on both sides of a tendon, and multipennate (where the tendon

is branched). For example, many muscles in our limbs have either a parallel or a fusiform (cigar-shaped, with the fascicles nearly parallel) arrangement. Sphincter muscles that close an opening, such as the mouth, have a circular fascicle arrangement, in which the fascicles encircle the opening. The closure muscle in the crab claw is bipennate. The fascicles are arranged in an angle on either side of centrally located tendons.

Fiber types

Fascicle arrangement is one way by which muscles can be tailored for a specific task. Another major difference between muscles used for different purposes is the properties of the individual muscle fibers. Most muscles contain more than one type of muscle fiber. In vertebrates, there are four main types of skeletal muscle fibers, each with a specific property. These four types of skeletal muscle fibers can be divided into two principal groups: tonic and phasic. Tonic muscle fibers occur in muscles that maintain the posture of animals such as amphibians, reptiles, and birds. They are also present in the muscles that move the eyeball in its socket. Tonic fibers contract slowly, and each fiber is able to produce a graded contraction. In contrast, a phasic fiber responds to a nerve

▶ LOWER FACIAL MUSCLES

The orbicularis oris muscle surrounds the mouth and is a type of sphincter muscle: the fascicles are arranged in a circle.

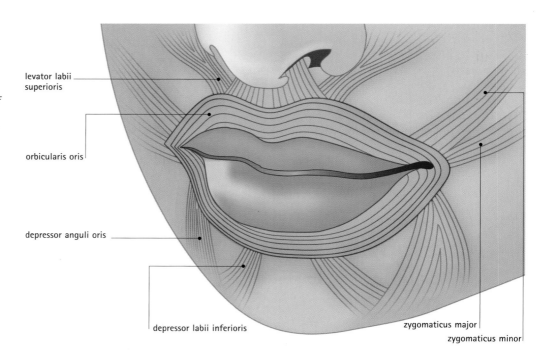

levator labii superioris

orbicularis oris

depressor anguli oris

depressor labii inferioris

zygomaticus major

zygomaticus minor

impulse in an all-or-nothing fashion: that is, either the fiber contracts with maximal force or it does not contract at all.

There are three types of phasic fibers: type I is slow oxidative (SO); type IIa is fast oxidative-glycolytic (FOG); and type IIb is fast glycolytic (FG). SO fibers are found in mammalian muscles that maintain body posture. They have a dark red appearance because they contain large amounts of the respiratory pigment myoglobin and many mitochondria. These enable the cells to generate ATP by aerobic processes. SO fibers contract slowly and also relax slowly.

FOG fibers contain fewer mitochondria and less myoglobin than SO fibers. That is because FOG fibers generate some ATP by anaerobic processes and some through aerobic respiration. They contract quickly and they also fatigue relatively slowly. FOG fibers are specialized for repetitive movements, such as walking and running, and they are especially abundant in the flight muscles of migratory birds such as ducks, seabirds, and warblers.

FG fibers contract very quickly, but they are also very quick to fatigue. They contain little myoglobin and few mitochondria, and they therefore have a much lighter appearance than the other types of fibers. In humans, these fibers are used for "burst activities," such as throwing a ball. Typically, skeletal muscles have a mixture of phasic fiber types, of which about half are SO fibers. The proportions vary depending on the role of the muscle.

During exercise, fast glycolytic fibers are engaged first, and the oxidative types of fiber take over for more prolonged or strenuous activity. In contrast to the mixture of fiber types in most vertebrate skeletal muscles, the muscle fiber types of swim muscles in many fish are anatomically separated from each other. In fish such as mackerel and tuna, which for most of the time cruise at relatively low

Force versus speed

When muscles contract, they pull two skeletal structures closer together. Often a muscle on the one side of a joint is pulling on a bone on the other side of the joint. The bone that is moved functions as a lever; and for levers there is a trade-off between the speed and force of the movement. Greater speed can be generated at the expense of reduced force if the insertion point of the muscle's attachment to the bone is close to the joint. Conversely, the strength of the movement can become greater if the insertion point is farther away from the joint, but in this case the potential for speed in the movement is reduced.

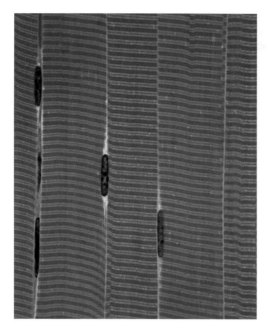

▲ *Typically, around half of an animal's skeletal muscle is made of SO fibers. Skeletal muscle has distinctive striations, and each cell has many nuclei. The dark stripes are A bands; between them are the I bands; and in the middle of the I bands are the Z lines.*

IN FOCUS

Mating call

Many species of bony (teleost) fish communicate with sounds. The mating call of the male oyster toadfish is said to sound like a steamboat whistle and is so loud that it can be heard from above the surface of the water. The sound is produced by muscles drumming on a gas-filled balloonlike structure, called the swim bladder. The teleost swim bladder is best known for functioning to keep a fish neutrally buoyant in the water; but it also has other roles, in hearing and sound production. To produce sound the muscles drum against the swim bladder at a very fast rate. In oyster toadfish the sound-producing muscles are able to contract and relax 200 times per second.

speeds, SO fibers are located in a band just under the skin, roughly along the lateral line. FG fibers, which in many species dominate the muscle mass, are located more deeply. Whereas the SO fibers are used for cruising, the FG fibers are recruited for fast swimming and for the escape response. The latter is the few forceful fin strokes that a fish makes to get away when startled. In squid, there is a similar anatomical differentiation between oxidative and glycolytic swim muscles.

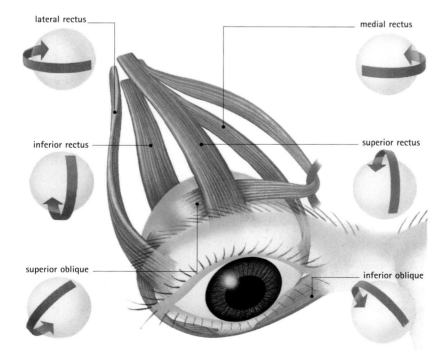

◄ EYE MUSCLES
The muscles around the eyes are made of tonic fibers. The contraction and relaxation of tonic fibers is much slower than that of the more common phasic fibers. Tonic fibers use less energy than phasic fibers and are thus useful where sustained contraction is necessary, such as keeping eyes open and moving throughout the day. Human eyes are moved by six main muscles.

Cardiac muscle

COMPARE the form of a vertebrate heart, such as that of a *HUMAN* or an *ELEPHANT* with that of an invertebrate, such as a *DRAGONFLY* or a *GIANT CLAM.*

CONNECTIONS

The vertebrate heart is made of cardiac muscle. Like skeletal muscle, cardiac muscle appears striated under a microscope. This is because it has regularly arranged sarcomeres, complete with the same bands and Z disks as the skeletal fibers. There are, however, notable differences between skeletal and cardiac muscles. While skeletal muscle fibers have the structure of long and parallel "cables" of fused cells, cardiac muscle cells are much shorter and have only one nucleus each.

A distinguishing feature of cardiac muscle cells is that they branch and are connected to neighboring muscle cells by intercalated disks. The intercalated disks contain proteins that keep the cells tightly joined at junctions called desmosomes as well as proteins that form pores

(called gap junctions) between neighboring cells. The gap junctions allow electric currents to be spread between muscle cells.

In addition to the muscle cells that are able to contract, the vertebrate heart has a special electrical conduit system that is made from modified muscle cells that cannot contract. The conducting fibers include separate bundles of conducting fibers that spontaneously generate rhythmic electric impulses. This is a feature that distinguishes vertebrate cardiac muscle from skeletal muscle, which requires nervous stimulation to contract. Another important difference between these two types of striated muscle is that the duration of each electrical impulse spread over the membranes of the muscle cells is 10 to 15

▶ HEARTBEAT REGULATION
Changes in the rate and strength of the heartbeat are regulated by signals from the autonomic nervous system (ANS), which is not under our conscious control. Information from sensory nerves travels to the regulatory center in the spinal cord. The heartbeat is then adjusted by two different branches of the ANS: the parasympathetic branch and the sympathetic branch.

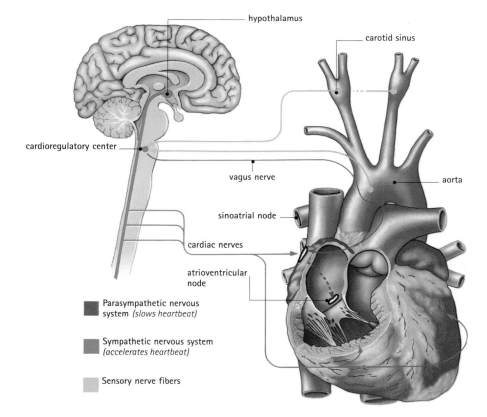

hypothalamus

carotid sinus

cardioregulatory center

vagus nerve

aorta

sinoatrial node

cardiac nerves

atrioventricular node

■ Parasympathetic nervous system *(slows heartbeat)*

■ Sympathetic nervous system *(accelerates heartbeat)*

■ Sensory nerve fibers

times longer in the cardiac muscle than in skeletal muscle. The result is that the contraction in each cardiac muscle cell is also of much longer duration than the contraction in the skeletal muscle fibers.

Nerve nodes

One of the two bundles of conducting fibers that can spontaneously discharge rhythmic electrical impulses is called the sinoatrial (SA) node. The SA node functions as the pacemaker for the heart and initiates the rhythmic contractions of the heart muscle of all mammals. In mammals, the SA node is located in the wall of the right atrium, and from there the electrical impulses are spread through pathways of conducting fibers over the two atria, resulting in the contraction of the right and left atria in succession.

The conduction pathways lead the electrical impulse to the second fiber bundle, called the atrioventricular (AV) node, which connects the atria to the ventricles electrically. From here, the conducting fibers direct the electrical signal along either side of the wall that separates the two ventricles down to the bottom of the ventricles. The conduction fibers then spread the electrical impulse back up along the walls of the ventricles. When each impulse arrives, the muscle fibers in its path contract, resulting in the squeezing of blood out of the respective heart chamber. The electrical impulse starts in the right atrium, and this is, consequently, the first chamber to contract quickly—followed by the left atrium and then, after a short delay, the ventricles. The brief pause between atrial and ventricular contractions is caused by a built-in delay in electrical conduction from the atria to the ventricles across the AV node. Because blood flows from the atria to the ventricles, this delay permits the atria to complete the filling of the ventricles before the latter contract to eject the blood out from the heart.

As a result of the spontaneous electric discharges by the pacemaker cells, vertebrate hearts contract rhythmically without the need for any stimulation by nerves. Even if the heart is surgically removed from the body, it will continue to beat for some time. If sufficiently supplied with oxygen and nutrients, hearts from many lower vertebrates can continue to

CLOSE-UP

Neurogenic heart

Crayfish, lobsters, and other decapods have neurogenic hearts. A heart is neurogenic if the stimulation to initiate each heartbeat comes from nervous tissue. In the lobster, each muscle cell is connected to a nerve fiber. The muscle cell contracts when it is stimulated by a nerve impulse. The rhythmic pattern of nerve impulses that directs the lobster heart is generated by a cardiac ganglion, which is an aggregation of nine nerve cells at the top of the heart. Five of these nerve cells make connections with the different heart cells, and the other four function as pacemakers that spontaneously generate a pattern of impulses. The pacemaker cells make contact with the nerve cells that supply the heart, and these direct contraction in individual heart cells.

beat for hours on their own. Hearts that generate their own rhythmic contractions are called myogenic. Vertebrate hearts do receive signals from nerves, but rather than initiating contractions (as in skeletal muscle) they modulate the rate and strength of the heartbeat. Mammalian hearts are supplied with nerves by nerve fibers belonging to two separate branches of the autonomic nervous system (the part of the nervous system that controls our bodies without our conscious involvement): the parasympathetic branch and the sympathetic branch.

▼ *This colored electron micrograph image of cardiac muscle clearly shows the circular mitochondria between pink muscle fibers. The thick transverse bands of the fibers are Z lines, regions with a denser arrangement of actin muscle filaments.*

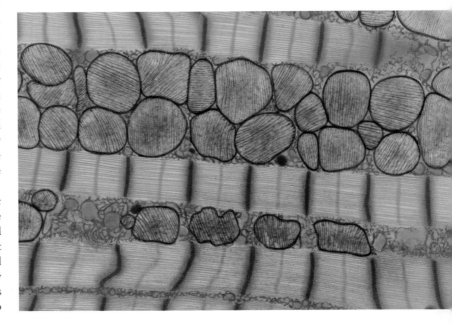

Smooth muscle

Smooth muscle cells are distinguished from striated muscle in that they lack internal arrangement into ordered sarcomeres; they are therefore not striated in appearance when viewed under a microscope. In vertebrates, smooth muscles are important for many of the functions in the body that are normally not under voluntary control. These include most muscles in the digestive system, the reproductive systems, the blood vessels, and the respiratory tract. Smooth muscle is also widespread in invertebrates, but invertebrate smooth muscle does not fall into the two functional groups present in vertebrates: single-unit and multiunit smooth muscles. However, even vertebrate smooth muscle is diverse and does not lend itself well to categorization.

Defining characteristics

Like cardiac muscle, but in contrast to skeletal muscle, each smooth muscle cell is an individual cell with one nucleus. Although sarcomeres are absent in smooth muscle, the contractile machinery is similar in smooth and striated muscle. Smooth muscle cells have myosin and actin filaments and many of the other proteins involved in contraction of striated muscle. Furthermore, as in striated muscle, contraction is achieved by the sliding of myosin and actin

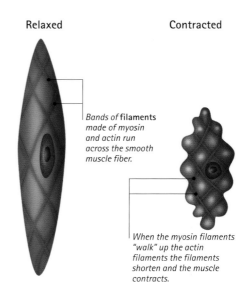

Relaxed   Contracted

Bands of filaments made of myosin and actin run across the smooth muscle fiber.

When the myosin filaments "walk" up the actin filaments the filaments shorten and the muscle contracts.

▲ SMOOTH MUSCLE FIBER
Smooth muscle fibers have one nucleus and are long when relaxed. When they contract, the crisscross arrangement of actin and myosin filaments causes the fiber to become compressed and globular.

filaments past each other as the myosin heads "climb" along the actin filaments. Instead of forming sarcomeres, the myosin and actin filaments of smooth muscle cells are organized into bundles close to the plasma membrane. The actin filaments are attached to dense bodies within the cytoplasm and to attachment plaques at the inside of the plasma membrane. These attachment sites are the functional equivalents of the Z disks in striated muscle.

Role of calcium ions

Smooth muscle cells have poorly developed sarcoplasmic reticuli and no T-tubules. As is the case with striated muscle, an increase in the Ca^{2+} concentration of the cytosol triggers contraction in smooth muscle. However, in smooth muscle much of the increase in cytosolic Ca^{2+} comes from the opening of Ca^{2+} channels in the plasma membrane, allowing Ca^{2+} to flow into the cell from outside. This flow results in a slow rise in the Ca^{2+} concentration of the cytosol

Bivalve catch muscle

In bivalves, such as clams and mussels, the two shells are kept closed by adductor muscles that are attached to either shell. When the creature is attacked by a predator, or is out of water during low tide, the adductor muscle keeps the two halves tightly closed, often for long periods of time. A regular muscle would tend to get tired and use up valuable energy conducting such a task. As a solution to these problems, the adductor muscles of clams and mussels are often a composite between striated fibers, which can deliver speed; and smooth muscle, which can provide endurance. In addition, the smooth muscle is capable of "locking" in a contracted state. No extra energy is required to maintain the muscle in the "locked" contracted position. In this way, these so-called catch muscles can keep the shells of the bivalve closed.

Two types of vertebrate smooth muscle

Vertebrate smooth muscles can be conveniently divided into single-unit and multiunit smooth muscles. In single-unit smooth muscle, all cells are connected to their neighbors by gap junctions, which allow the muscle cells to contract as a unit. As in the cardiac muscle, electric currents can be conducted across these gap junctions to induce contraction in the entire tissue. Also, like the cardiac muscle, single-unit smooth muscles have spontaneous activity. In contrast, multiunit smooth muscle cells have few gap junctions and operate individually when stimulated. They often have an extensive supply of nerves, and they can be activated or relaxed by nerves and hormones.

▶ *Single-unit smooth muscle is the most common form of smooth muscle. Nerves stimulate the single-unit muscle cells to contract by releasing neurotransmitters. The signal passes from one muscle cell to another through gap junctions, and the cells therefore contract as a unit. Multiunit smooth muscles are found in the uterus, the male reproductive tract, and in the eye. In multiunit muscle cells, each cell must be stimulated individually by neurotransmitters.*

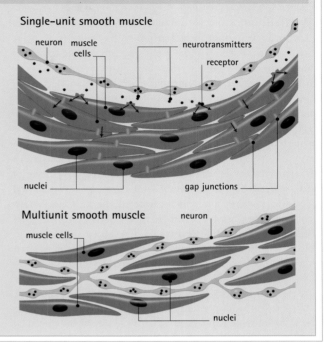

Single-unit smooth muscle

neuron   muscle cells   neurotransmitters   receptor   nuclei   gap junctions

Multiunit smooth muscle   neuron   muscle cells   nuclei

and, when stimulation ceases, a slow removal of Ca^{2+} from the cytosol. Consequently, most smooth muscles contract and relax very slowly.

Smooth muscles do not have troponin. Instead, there are several other pathways that link an elevated Ca^{2+} concentration to myosin–actin cross-bridge formation. These pathways have been discovered mainly from chicken's gizzard. Vertebrates have at least six types of smooth muscles, including vascular and respiratory. Invertebrates have many other types. It is still not clear how pathways for contraction differ in different types of smooth muscles or animals. The current model is a combination of what is known from various smooth-muscle types and animals.

In smooth muscle, a rise in the cytoplasmic concentration of unbound Ca^{2+} is sensed by a protein called calmodulin. This protein removes an inhibitory protein from actin, exposing the myosin-binding sites on the actin filament. At the same time, calcium-bound calmodulin activates an enzyme that enables myosin to form cross bridges with actin and contract.

There is also evidence that binding of Ca^{2+} directly to myosin light-chain proteins induces a change in shape that allows binding of the myosin head to actin. Other mechanisms involve binding of Ca^{2+} to a calcium-dependent enzyme, called protein kinase. This enzyme is present in many cells, where it controls different processes, but in smooth muscle it can regulate contraction and relaxation according to the concentration of Ca^{2+} in the cytosol.

Unlike skeletal muscle, which contracts in response to nerve signals only, contraction or relaxation in smooth muscle is regulated by numerous factors, such as nerve signals and hormones. Furthermore, smooth muscle is often supplied with nerves from several different nerve fibers, each of which may have a distinct influence on the contraction state of the muscle.

CHRISTER HOGSTRAND

FURTHER READING AND RESEARCH
Muscolino, Joseph E. 2005. *The Muscular System Manual: The Skeletal Muscles of the Human Body, 2nd Edition.* CV Mosby: St Louis, MO.

Nervous system

All organisms receive information, process it, and then produce an appropriate response. In complex animals, these functions are, for the most part, performed by two interconnected systems: the nervous system, and the endocrine system, which produces hormones.

The nervous system consists of often huge networks of nerve cells that perform three interconnecting functions. First, the nervous system allows animals to sense what is happening in their environment. Their environment is both internal (within the body, including such things as hormonal levels or the amount of stretch in a muscle) and external (outside the body—for example, as monitored by vision, hearing, and touch). Second, the nervous system processes this sensory information and compares the information from different senses. The processing can be a relatively simple reflex or extremely complex, as in human speech. Third, the nervous system enables animals to do things, primarily by controlling muscles and glands. The three functions can be accomplished amazingly quickly, within a few milliseconds (thousandths of a second). This speed of information transmission is achieved by electrical and chemical signals within and between nerve cells.

Processing the vast amount of sensory information and producing an appropriate response require nervous systems to be remarkably complicated. This complexity can be seen at all levels. For example, how individual molecules of the nervous system operate is a scientific field in itself (called biophysics). Most nervous systems have huge numbers of nerve cells and even more connections between them. Such complexity is necessary to produce the subtle range of behaviors most animals display.

Different types of nervous systems

The first nervous systems to evolve were probably networks of identical nerve cells that connected different parts of the body and allowed a general response to limited sensory input. Later, nerve cells specialized and formed physical groups for specific functions. Such groups are called ganglia (singular, ganglion). For example, in segmented animals, such as earthworms, there are ganglia in each segment. In gastropod mollusks, such as slugs and snails, the nervous system includes a buccal (mouth) ganglion, which helps control feeding; and a pedal ganglion, which is involved in locomotion. Sponges are the only multicellular animals that do not possess a nervous system.

Most animals tend to move mostly in one direction. The front end (the head of the animal) tends to encounter

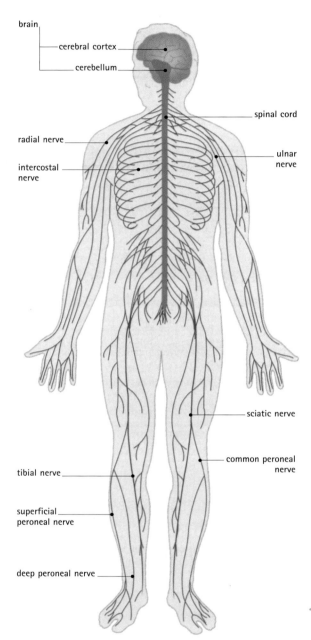

▲ NERVOUS SYSTEM
Human
Typical of vertebrates, humans have a nervous system made up of a central nervous system (the brain and spinal cord, shown in orange) and the peripheral nervous system (a network of nerve fibers branching to every part of the body, shown in blue).

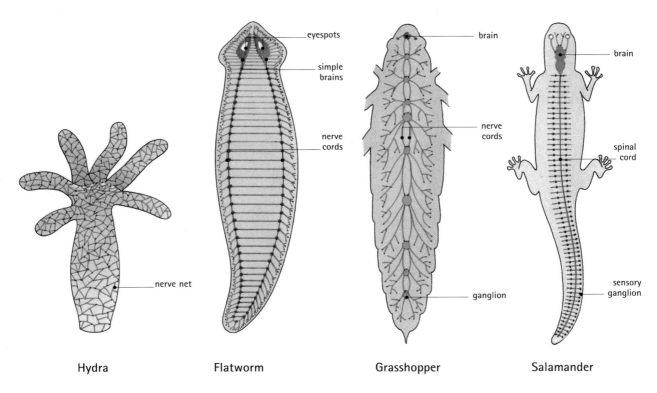

Hydra   Flatworm   Grasshopper   Salamander

▲ *From left to right, these animals have a progressively more complex nervous system, from a hydra with a loose nerve net to a salamander with a brain and spinal cord.*

HOW THE NERVOUS SYSTEM WORKS The nervous system is made up of nerve cells (neurons). There are various anatomical types of neurons to match their specialized functions. Neurons deal with information—receiving it, transmitting it, and relaying it to other neurons, muscles, or glands—by using electrical and chemical signals. *See pages 110–113.*

SENSORY AND MOTOR SYSTEMS Sensory systems allow animals to see, hear, feel, taste, and smell. There are other senses as well: some animals can detect electrical or magnetic fields. Motor systems allow animals to make appropriate responses based on sensory input. *See pages 114–119.*

INVERTEBRATE NERVOUS SYSTEMS An extraordinary diversity of nervous systems—as diverse as the group itself—occurs in invertebrates. *See pages 120–121.*

VERTEBRATE NERVOUS SYSTEMS Vertebrates have a central nervous system—consisting of a brain and spinal cord—and a peripheral nervous system that connects to sensory organs and muscles. *See pages 122–127.*

stimuli first, and most sensory structures are of greatest use there. All this sensory input is therefore dealt with by the frontmost ganglion, so it has evolved to be larger and have more nerve cells than the others. This large front ganglion is often called the brain. Brains of one type or another are found in both invertebrates and vertebrates. So, in most animals there is some form of brain at the front end and a nerve cord or cords (with ganglia) running the length of the body that collect sensory information and distribute motor commands locally. Within this basic form, considerable variations occur across the animal kingdom. For example, in insects two parallel nerve cords run along the belly, whereas in vertebrates a single cord runs along the back.

The nervous system of vertebrates may be divided into two main parts—the central nervous system (CNS), which consists of the brain and spinal cord; and the peripheral nervous system, which comprises the nerves that connect with the brain and spinal cord.

All nervous systems are complicated, and the largest in terms of numbers of nerve cells (about 100 billion in the human brain) and the hardest to understand is the human nervous system. Unraveling the workings of the human brain, which is much more complicated than even the most sophisticated computer, remains one of the greatest challenges to scientists.

How the nervous system works

CONNECTIONS

COMPARE the nervous system with the **ENDOCRINE AND EXOCRINE SYSTEMS**. The nervous and endocrine systems work closely together and interact strongly. Nerve cell activity can control hormone release, and in turn hormones can affect the nervous system. The endocrine system usually operates more slowly (from seconds to years) than the nervous system, which generally takes fractions of a second to work.

The word *neuron* is another name for a nerve cell. Neurons have the same basic anatomy as other body cells. They have a nucleus containing genetic material (DNA), have cytoplasm with organelles, and are surrounded by a cell membrane. Some features are common to most neurons. The nucleus is contained within a cell body, or soma. There are a number of branching structures called dendrites, which receive information from other neurons; and one or more long filament-like extensions called axons, which send information. A nerve is simply a bundle of axons that serve the same part of the body.

Neurons perform many tasks within the nervous system, so their anatomy is amazingly varied to match their specialized functions.

Because they have to send information over distances, some neurons are the longest cells in the body and can be more than 3 feet (1 m) long in humans. Neurons can be broadly divided into three classes. Sensory neurons receive and then transmit information from sense organs and cells. Motor neurons send the output of the nervous system to muscles and glands. Interneurons pass information between neurons and are the most common type of neuron in the nervous system.

For the nervous system to work, neurons need to receive, process, and pass on information. This capability requires a means of sending signals over long distances within individual neurons, along their axon. It also needs a way for signals to pass between neurons.

▶ **RESPONDING TO STIMULI**

In vertebrates, stimuli detected both outside and inside the body produce information that is transmitted by sensory neurons to the central nervous system. There, the information is interpreted, bringing about an appropriate response, by motor, or efferent, neurons.

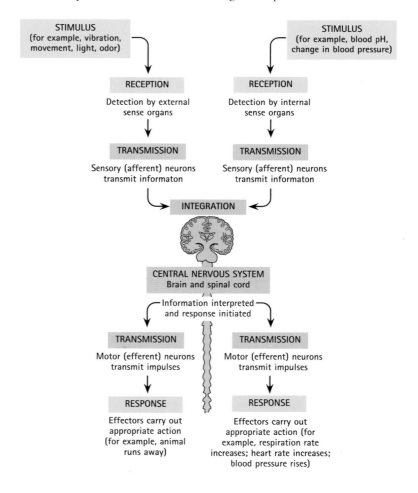

STIMULUS (for example, vibration, movement, light, odor)

STIMULUS (for example, blood pH, change in blood pressure)

RECEPTION — Detection by external sense organs

RECEPTION — Detection by internal sense organs

TRANSMISSION — Sensory (afferent) neurons transmit informaton

TRANSMISSION — Sensory (afferent) neurons transmit informaton

INTEGRATION

CENTRAL NERVOUS SYSTEM — Brain and spinal cord

Information interpreted and response initiated

TRANSMISSION — Motor (efferent) neurons transmit impulses

TRANSMISSION — Motor (efferent) neurons transmit impulses

RESPONSE — Effectors carry out appropriate action (for example, animal runs away)

RESPONSE — Effectors carry out appropriate action (for example, respiration rate increases; heart rate increases; blood pressure rises)

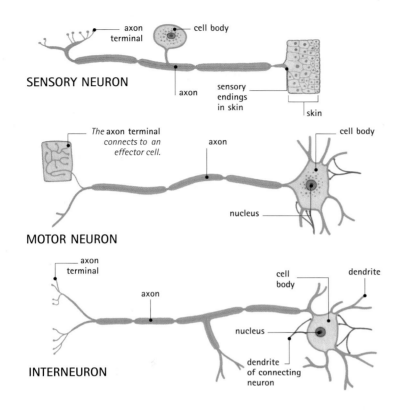

SENSORY NEURON

axon terminal — cell body

axon — sensory endings in skin

skin

MOTOR NEURON

The axon terminal connects to an effector cell.

axon

cell body

nucleus

INTERNEURON

axon terminal

axon

cell body — dendrite

nucleus

dendrite of connecting neuron

How signals are transmitted

The concentrations of electrically charged ions vary on either side of a neuron's membrane, creating a voltage difference across the cell membrane. At rest, this voltage difference is called the resting membrane potential and is about −70 millivolts (mV). In other words, the inside of the neuron is 70 mV more negative than the outside, and the membrane is said to be "polarized." This difference results because there is a higher concentration of positively charged sodium ions outside the cell, and inside the cell a higher concentration of negatively charged chloride ions and a lower concentration of positively charged potassium ions. Protein molecules in the cell membrane maintain this difference by actively pumping ions from one side of the membrane to the other.

Electric signals are transmitted in neurons when there is a change in the membrane potential—that is, when there is a movement of ions across the cell membrane. The most common form of transmission is called an action potential, or nerve impulse. If a stimulus causes the membrane potential to become less negative

▲ THREE TYPES OF NEURONS

Sensory neurons detect and transmit sensory information. Motor neurons carry impulses to effector cells such as muscles. Interneurons pass information between neurons.

Squid giant axons

Many early studies on neuronal properties used neurons that mediate the rapid escape response in squid. These neurons have axons that are very large—up to 0.04 inch (1 mm) in diameter. Because of their large size, isolated squid axons can have metal electrodes inserted into them to measure the membrane potential and how it responds to different stimuli. The contents of the axon can also be squeezed out, like toothpaste from a tube. This allows the concentrations of various ions within the axon to be measured and compared with concentrations outside the axon.

(depolarized), an action potential is triggered. That happens, however, only if the stimulus is strong enough to breach a certain threshold.

At the start of an action potential, ion channels in part of the membrane allow an inward rush of sodium ions. The local membrane potential momentarily becomes positive, passing 0 mV. The sodium channels then begin to close, and potassium ions followed by sodium ions are pumped out of the cell. Then, there is a rapid return to the resting membrane potential at the site of depolarization. The whole process of generating an action potential takes only 2 to 5 milliseconds.

When a region of cell membrane triggers an action potential, it creates a stimulus that depolarizes a neighboring region of membrane. In this way, the action potential is propagated,

Venoms that attack the nervous system

Venoms often contain specific molecules that act against the nervous system. Some of these molecules (such as tetrodotoxin from pufferfish) stop neurons from producing action potentials. Others (like bungarotoxin from snakes called kraits) block the receptor molecules at synapses (junctions between neurons) and stop communication between neurons. For victims of the venom, this can have painful or even lethal consequences. However, when isolated, these molecules can be used in experiments to study the nervous system. By using venom molecules to block a specific activity, scientists can investigate its normal function.

Electric plants

Neurons are not the only cells that are electrically excitable. Muscle cells also have a resting potential that responds to electrical input from neurons. In addition, some plant cells have resting potentials and can produce the equivalent of action potentials—for example, the cells involved in the fast movement with which the insectivorous Venus flytrap captures its prey.

or travels, along the neuron—usually starting at the cell body and then along the axon—at up to 200 miles (320 km) an hour. Action potentials are all-or-nothing, so they cannot vary in size. Information is therefore transmitted by varying the frequency and timing of action potentials.

How one neuron connects with another

For the nervous system to be efficient, signals need to pass between neurons. Information is passed between neurons at junctions called synapses. At most synapses, the axon of one neuron makes one or more contacts with the dendrites or cell body of another. There may be more than 1,000 synapses on the cell body and

▶ **NERVE IMPULSE**
At rest, there is a higher concentration of negatively charged ions inside a neuron than outside, creating a voltage potential (difference) across the membrane (1). A stimulus above a certain threshold level depolarizes a region of cell membrane and triggers an action potential, in which there is a local influx of sodium ions into the cell (2). Sodium and potassium ions are then pumped out of the axon at the site of depolarization, and the membrane returns to the resting potential; however, the neighboring region of membrane is then sufficiently depolarized to trigger another influx of sodium ions; and in this way the action potential is transmitted along the axon (3).

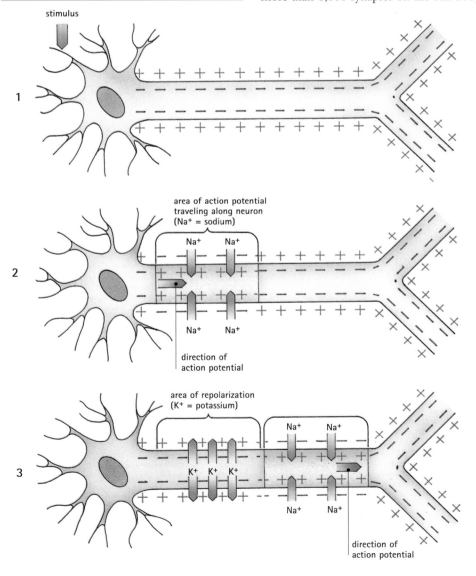

IN FOCUS

Holding it all together

Neurons cannot work on their own. They are assisted by a range of cells called glial (Greek for "glue") cells. Glial cells have various roles, from physical support to electrical insulation. Although (unlike neurons) they cannot signal rapidly over long distances, glial cells are vital for the nervous system to work. In fact, glial cells are 10 to 50 times more numerous than neurons. Some glial cells even provide growth factors, which are chemicals essential for some neurons to survive. One important type of glial cell in vertebrates is the Schwann cell. Schwann cells make a form of protective sheath around axons. However, there are small uninsulated gaps at regular intervals. There, the action potential jumps from one gap to the next, allowing it to travel much faster.

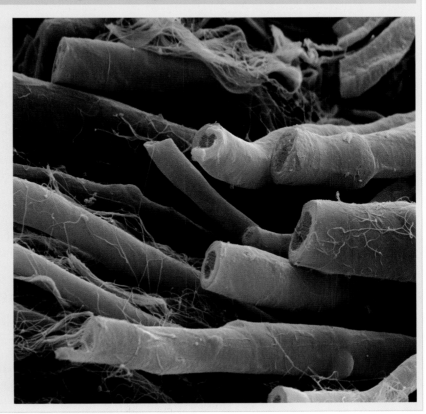

▶ *This scanning electron micrograph shows axons surrounded by a protective sheath (purple) that is formed of concentric layers of Schwann cells.*

dendrites of just one human motor neuron. The most common type of synapse is the chemical synapse. The neuron making contact has a number of vesicles, which are internal spheres of membrane that contain molecules of a chemical messenger called a neurotransmitter. A neurotransmitter is a specific molecule that transfers a signal between neurons. Two examples of neurotransmitters are dopamine and acetylcholine. When an action potential arrives at a synapse, the vesicles fuse with the cell membrane and release the neurotransmitter into the synaptic cleft, the tiny gap between a neuron and its target cell. The neurotransmitter molecules diffuse rapidly across the narrow cleft and then bind to receptor molecules on the cell membrane of the receiving neuron. Receptor molecules are specific to individual neurotransmitters and can

work in a variety of ways. Some receptors open or close channels in the membrane, allowing certain electrically charged ions to enter or leave the neuron. Others work through chemical signaling systems within the neuron. The final result is a change in the membrane potential, either positive or negative, of the receiving neuron. Individual neurons usually receive information from numerous synapses. If the sum of the inputs depolarizes the neuron sufficiently, it will trigger one or more action potentials, passing on an electrical signal.

At an electrical synapse the two neurons make a direct electrical contact. Transmission at an electrical synapse is fast, but there is not much scope for modifying the signal, as there is at a chemical synapse. Modifying signals at chemical synapses is necessary for processes like learning and memory.

Sensory and motor systems

Sensory systems allow animals to detect conditions inside and outside the body, in addition to any changes in those conditions. To detect such information, energy from the sensory stimulus, such as photons of light or vibrations in the air, or even more direct stimuli such as touch, must be converted into a signal the nervous system can distribute and process. Sensory information is first detected in cells called sensory receptors. Receptors are usually modified neurons or epithelial cells on the outer surface of the body and the walls of internal cavities. Receptor cells transform the energy of the sensory stimulus into an electrical signal. This electrical signal is then transmitted to the CNS as a series of impulses called action potentials, initiated either in the receptor or in neurons connected to it.

Receptor cells will detect sensory signals only within a certain range. For example, humans can detect light only between the wavelengths of 380 and 760 nanometers (0.00000038 and 0.00000076 m) and sound between frequencies of 20 and 20,000 hertz (Hz, air vibrations per second). Different animals can have different sensory ranges. Bats can hear sound frequencies as high as 120,000 Hz (ultrasound), whereas elephants can hear frequencies as low as 1 Hz (infrasound). Often, different receptor types are used to detect different ranges of a sensory stimulus. In the human eye there are three types of receptors responsible for color vision, which are sensitive to red, green, or blue wavelengths of light. The brain compares the information from these different receptors to achieve a full perception of color. Receptors can also register the strength (intensity) of a sensory stimulus. A more intense stimulus results in more action potentials transmitted close together, and in more receptors becoming active.

Sensory processing
Although the sensory stimulus may vary (touch, pain, light, sound), the final input received by the nervous system is always a series of action potentials. This sensory signal is then converted into a perception of a sense in the brain. So you do not really "see with your eyes," but in fact form a perception of vision in your brain. Transforming a series of action potentials into a perception of the environment starts at the receptor cell itself. When a change in the environment occurs, a strong neuronal signal is usually generated (many, frequent action potentials). If the environment then maintains this new configuration, the signal will reduce over time. The number and frequency of action potentials will be reduced. This is a process known as sensory adaptation. Note that this is quite separate from the meaning of adaptation in an evolutionary sense. One example of sensory adaptation is putting on a wristwatch: when you first put a watch on your wrist you can feel it, but you become less aware of it over time. The advantage of adaptation is

▼ *This false-color scanning electron micrograph shows photoreceptors in the human retina. The tall brown structures are rods (responsible for vision in dim light), and the less common, short orange structures are cones (responsible for color vision). Magnified 3,300 times.*

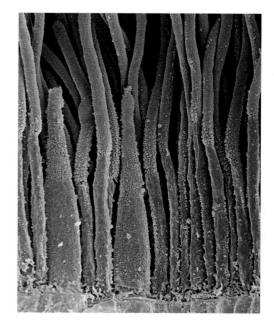

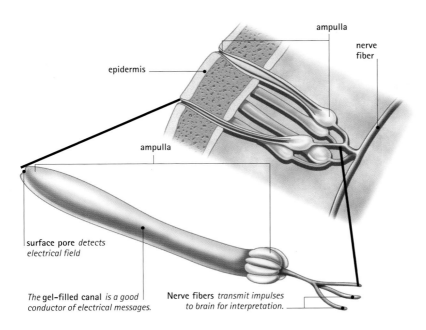

ampulla

nerve fiber

epidermis

ampulla

surface pore *detects electrical field*

The **gel–filled canal** *is a good conductor of electrical messages.*

Nerve fibers *transmit impulses to brain for interpretation.*

◀ **AMPULLAE OF LORENZINI Hammerhead shark**
Sharks and rays have electroreceptors called ampullae of Lorenzini embedded in their head. These receptors allow them to detect the electrical field of other aquatic animals, including prey.

that it provides the organism with information about changes in the environment without overloading the nervous system. Much more sophisticated sensory processing occurs in the CNS. This allows us, for example, to interpret different wavelengths of light as different colors. The higher processing and integration of sensory information is a complex issue and one that is only just starting to be understood.

Sensory receptor cells can occur singly or in large groups. Receptors are usually part of a larger structure called a sensory organ. Sensory organs range from simple individual spines on a cockroach leg, which detect touch, to complex structures such as the vertebrate eye or ear. Sensory organs can concentrate and amplify a sensory signal. The bones of the human inner ear amplify sound waves by 20 times before they reach the receptors. Sensory organs can also allow an animal to detect which direction a sensory signal is coming from.

Sensing conditions within the body

The sensory systems in vertebrates send information to the nervous system, which then adjusts conditions, usually without conscious control. Mechanoreceptors, for example, are sensory neurons that detect the degree of muscle contraction and the position of the skeletal system. The nervous system processes the information and sends signals to the

muscles to maintain balance when we are standing or walking. Sometimes, we are more aware of the conditions within the body. For example, sensory receptors in the mammalian brain may detect an increasing concentration of substances such as salt and sugar in the bloodstream. This signal will be processed by the brain, which then stimulates a sense of thirst. After the mammal has a drink of water, the concentration of salt and sugar in the blood decreases and the feeling of thirst diminishes.

IN FOCUS

Sensory transduction

The method by which a receptor converts a sensory input to an electrical signal is called sensory transduction. For example, the sense of smell relies on specific chemicals binding with protein molecules in the cell membrane of an olfactory receptor cell. The binding causes ion channels in the membrane to open, allowing electrically charged ions into, or out of, the receptor cell. The movement of charged ions alters the receptor cell's resting potential, which can initiate action potentials. These impulses then travel along nerves to the brain, where the relevant smell is perceived.

Cephalopod and vertebrate eyes

The visual system is extremely useful for most animals, since it allows a great deal of information to be detected, often at long distances. For this reason, eyes are found in several groups. Eyes in cephalopods, such as octopuses and squid, and vertebrates are remarkably similar in many ways. Both types of eyes are spherical, with an outer layer (cornea) and a small area where light can enter (the pupil). In both cases, the diameter of the pupil varies according to the light intensity. Both types also have a lens to help focus the light onto a sensory surface (the retina). In each case, the eye has a similar function and needs the same high performance levels. The similarity of eyes in cephalopods and vertebrates is an example of convergent evolution—structures with different origins that have evolved to become alike.

▼ EYE
Human
Light rays are focused by the cornea and lens to produce an upside-down image on the retina, the light-sensitive tissue lining the back of the eye. Electrical signals from stimulated cells in the retina travel to the brain via the optic nerve for interpretation.

Sensing the outside world

Nervous systems can extract a wide variety of information about the outside world. The types of information can be broadly divided by the form of input that is being detected.

Mechanoreceptors detect mechanical energy. This mechanical sense can be direct, such as touch and pressure; or indirect, as when an ear detects high-frequency vibrations in the air that are perceived as sound. The sense of balance

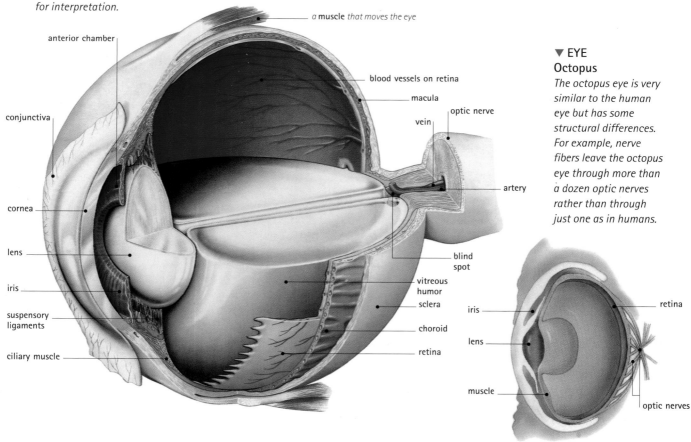

a **muscle** *that moves the eye*

anterior chamber

conjunctiva

cornea

lens

iris

suspensory ligaments

ciliary muscle

blood vessels on retina

macula

vein

optic nerve

artery

blind spot

vitreous humor

sclera

choroid

retina

▼ EYE
Octopus
The octopus eye is very similar to the human eye but has some structural differences. For example, nerve fibers leave the octopus eye through more than a dozen optic nerves rather than through just one as in humans.

iris

lens

muscle

retina

optic nerves

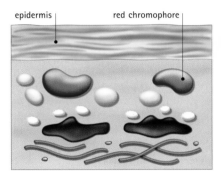

epidermis | red chromophore

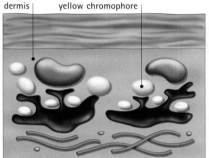

dermis | yellow chromophore

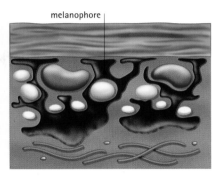

melanophore

▲ **COLOR CHANGE IN LIZARDS**
Chameleon, gecko, or agama
In some lizards, the ability to change skin color is controlled by the nervous system.

When red and yellow pigment cells called chromatophores dominate near the surface of the dermis, the lizard appears orange. As black and brown cells called melanophores

move up the dermis, the lizard appears terra-cotta. As the melanophores spread out under the epidermis, the lizard darkens to chocolate brown.

in mammals is also a form of mechanoreception as it relies on the movement (due to gravity) of hairlike projections on receptors in the vestibular apparatus of the inner ear.

Nociceptors detect pain. Different types of nociceptor respond to different causes of pain: extreme heat, pressure, chemicals (such as acid), or inflammation. Pain is an important sense, since it can result in avoidance of a damaging contact. Also, a painful damaged part of the body is more likely to be rested, speeding healing.

Photoreceptors detect light. The sensory range of visual systems can be very large in different animals and is often well outside the human visual spectrum. Insects, birds, and fish can detect ultraviolet light (wavelengths that are shorter than 400 nm). Some snakes have

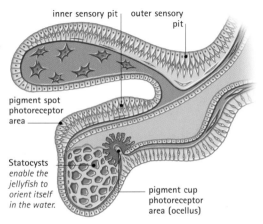

inner sensory pit | outer sensory pit

pigment spot photoreceptor area

Statocysts *enable the jellyfish to orient itself in the water.*

pigment cup photoreceptor area (ocellus)

◄ **NERVOUS SYSTEM**
Jellyfish
Around the edge of a jellyfish's bell are centers called rhopalia, which contain various sensory areas that enable the animal to detect light and chemicals and to detect the orientation of the bell with respect to gravity.

EVOLUTION

Losing your senses

Animals that are not exposed to a certain type of sensory input have often evolved over millennia in such a way that they lose the ability to detect this input. Dispensing with an unnecessary sense organ saves metabolic energy. This can be seen in certain animals, including some fish, that live in permanent darkness in caves and have lost their ability to see. Extreme cases can be seen in adult parasites, such as the tapeworm, that live inside warm-blooded animals. In these circumstances, the conditions are so constant that the parasites have very few sensory structures.

► *The yellow clumps in this colored micrograph of the human inner ear are hairlike stereocilia, which project from the ends of sensory cells. The inner ear converts sound waves into nerve impulses by stimulating the stereocilia, and these impulses travel to the brain for interpretation. Magnified 2,500 times.*

Hearing with legs and tasting with feet

Exactly where particular types of sensory information are detected varies in different animals. Sound is very important for crickets—for example, in finding a mate. To tell where a sound is coming from, the "ears" need to be as far apart as possible. The farthest distance possible on a small insect is between the legs, so crickets have tympanic membranes on their legs, which they use to hear mating calls. Flies need to detect food substances when they first land on a surface, so they have chemoreceptors on, and can "taste" with, their feet.

Chemoreceptors can detect both simple and complex chemicals. For example, the ability to taste salt (sodium chloride) is brought about by chemoreceptors on the tongue. Olfaction, the sense of smell, is another form of chemoreception. Most chemoreceptors respond to general groups of structurally similar molecules. Sometimes the receptor is highly sensitive to just one chemical of particular importance to an animal. Male moths often have chemoreceptors in their antennae, which are particularly sensitive to sex pheromones released by female moths.

Electroreceptors can detect electrical fields and exist in animals as diverse as sharks and platypuses. Electroreceptors are used to detect the small electrical fields around prey items buried in sand and mud, where other senses cannot penetrate.

There is good evidence to suggest that some animals (such as pigeons) can detect the earth's

receptors capable of detecting infrared radiation, which they use to detect the body heat of their prey. Honeybees can see the polarization of light in the sky, and they use it to navigate. For many animals, the way in which they perceive the environment must be very different from a human's point of view.

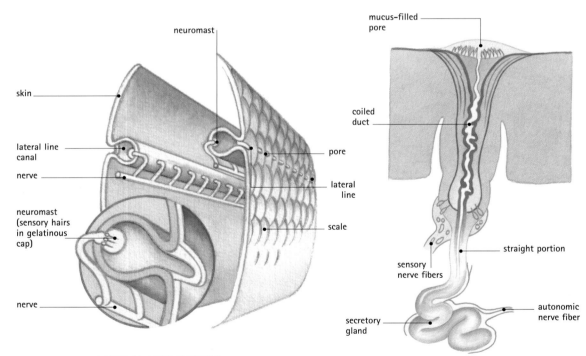

▲ LATERAL LINE SYSTEM
Trout
The lateral line system is a sensory organ that runs around the head and down the sides of a fish. Sensory hairs in structures called neuromasts detect movements in water, allowing the fish to sense not only its prey but also its predators.

▲ ELECTRORECEPTOR
Platypus
There are about 10,000 mucus-filled electroreceptor glands in the platypus's bill. These receptors help the animal locate prey underwater or even buried in mud where they cannot be seen.

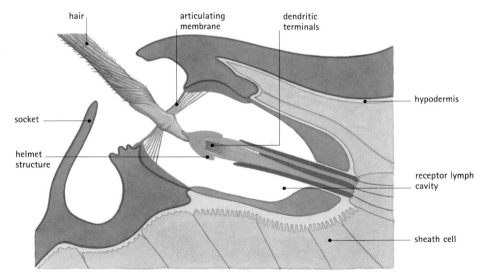

▶ **TRICHOBOTHRIUM**
Spider
Trichobothria are long, fine hairs arising from sockets on the spiders' legs. These hairs are mechanoreceptors, which detect air vibrations and currents.

hair

articulating membrane

dendritic terminals

hypodermis

socket

helmet structure

receptor lymph cavity

sheath cell

magnetic field and use it for navigation. The brains of pigeons contain magnetite, a naturally magnetic mineral. However the exact location, anatomy, and physiology of the possible magnetoreceptors have yet to be found.

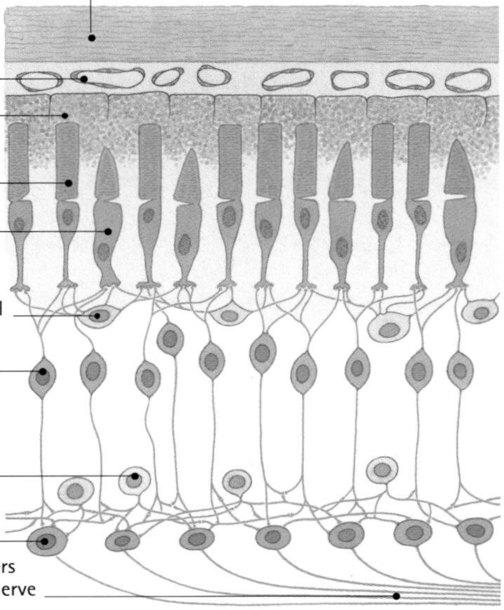

Motor systems: Effectors

The output of the nervous system travels to organs called effectors. By far the most common types of effectors are muscle cells of various types. Motor neurons make synapses with muscle cells; and when the neurons are active, the muscle contracts. In addition to muscles, there are other types of effectors. One main group comprises the hormone-producing glands of the endocrine and exocrine systems, in which neuronal activity can stimulate or stop the release of substances. For example, in mammals the hypothalamus, at the base of the brain, can signal the pituitary gland to release a variety of hormones into the bloodstream.

sclera

choroid layer

pigment layer

rod

cone

horizontal cell

bipolar cell

amacrine cell

ganglion cell

nerve fibers of optic nerve

▲ **CROSS SECTION OF RETINA**
Mammalian
Rods and cones produce electrical signals when light strikes the retina. These signals are modified by other cell types, such as bipolar cells, before leaving the retina via the optic nerve.

▼ *This light micrograph shows a neuromuscular junction, where a nerve cell connects to muscle. Four ends of an axon branch onto muscle cells, terminating in synapses (black dots). Magnified 300 times.*

Invertebrate nervous systems

Invertebrates possess a dazzling array of different nervous systems. Even within a single taxonomic group, there is much variation. This can be demonstrated by looking at the mollusks. Some mollusks, such as clams, have just a few widely spaced and similarly sized ganglia. No specific coordinating brain is needed for an adult life spent in one place. Octopuses, on the other hand, have a very sophisticated nervous system with a relatively large brain. This complexity reflects their active hunting behavior and excellent learning and memory abilities.

Nerve nets to nerve cords

Among the simplest nervous systems is that of cnidarians called hydras. (Cnidarians include corals, sea anemones, and jellyfish.) A hydra's neurons are arranged in a netlike manner. Since these nerve nets lack ganglia, there is no central control area, and the processing of information is distributed throughout the nervous system.

Most of the synapses between neurons are electrical, and action potentials can travel in either direction along the axon. This structure enables a stimulus at any point to trigger action potentials throughout the nerve net and lead to a generalized response. Other cnidarians show some degree of grouping of neurons, particularly around the mouth, where sensory information is most varied. Some planarians (flatworms) possess a network of nerves throughout the body, and they have a distinct brain in the head region. Other planarians have two nerve cords extending along the body, connected and extended by a series of lateral nerves to form a ladderlike system.

Complex invertebrate nervous systems

Even an animal as apparently simple as the earthworm (an annelid) has a complex nervous system. Earthworms possess a brain and a nerve cord along the body, with a ganglion in each segment. Each ganglion sends lateral nerves

▶ **Giant clam**
The giant clam does not have a true brain. Instead, it has three pairs of interconnected ganglia. These are the cerebropleural ganglia, which send nerve fibers to the the palps (feeding appendages), mantle (body), and anterior adductor muscle; the visceral ganglia, which send nerve fibers to the heart, mantle, posterior adductor muscle, gut, gills, and siphon; and the pedal ganglia, which send nerve fibers to the foot.

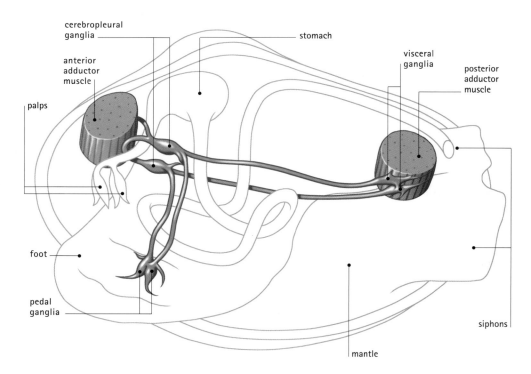

throughout the segment to gather sensory information and distribute motor commands. This basic pattern is also found in many other invertebrate groups, including the arthropods (such as insects and crustaceans). In arthropods, the segmental ganglia have often evolved to be physically fused, so their nervous system is more centralized and specialization of groups of segments is possible.

A very different nervous system is seen in the echinoderms (starfish and sea urchins). These have a radial nervous system with nothing resembling a central brain. There is a nerve ring around the mouth and nerves extending into each arm. Sometimes there are small ganglia associated with the tube feet. The unique nervous system of echinoderms allows any radial portion of the body to act as the "head." This feature also allows the nervous system to be regenerated following even major injury, such as the loss of a limb in a starfish.

▶ **Earthworm**
This worm has a brain and a nerve cord running along its body. Within each segment, a ganglion gives rise to three main nerves (one is shown here), which divide further.

COMPARATIVE ANATOMY

Naming neurons

Some invertebrate species have specific neurons that can be identified from one individual animal to the next. The neurons can then be named (or numbered) and their functions studied. Some of the specific neurons control particular aspects of a behavior. For example, artificially stimulating action potentials in a neuron called CBI-2 in the sea hare (a type of sea slug) induces feeding rhythms—as though the animal is eating. In most vertebrates, the neurons are so small and numerous that this kind of identification and naming is not possible. One exception is the Mauthner cell in bony fish, which induces an escape response when it is stimulated.

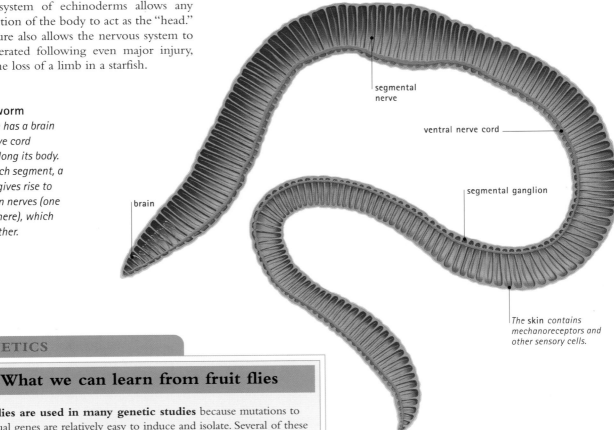

brain

segmental nerve

ventral nerve cord

segmental ganglion

The skin contains mechanoreceptors and other sensory cells.

GENETICS

What we can learn from fruit flies

Fruit flies are used in many genetic studies because mutations to individual genes are relatively easy to induce and isolate. Several of these mutations have involved changes to nervous system function. Mutant fruit-fly genes called dunce and rutabaga result in poor learning. The mutations in these cases affect enzymes involved in sending messages within neurons. Neuroscientists can use these mutations to study how learning is achieved at the cellular level.

Vertebrate nervous systems

Vertebrate nervous systems are not as varied as those of invertebrates. However, vertebrate nervous systems are generally capable of much higher levels of processing and interpreting information. Vertebrate nervous systems also typically generate more flexible and sophisticated behaviors. This is particularly so in birds and mammals and especially in the human nervous system. The vertebrate nervous system can be broadly divided into two parts: the peripheral nervous system (PNS) and the central nervous system (CNS). The CNS consists of the brain and spinal cord, and the PNS is made up of the nerves connecting the CNS with the rest of the body, including the sense organs and muscles.

Peripheral nervous system

Anatomically, the PNS consists of a series of nerves extending from the brain (the cranial nerves) and spinal cord (the spinal nerves). The PNS has two main functions. It collects sensory information and distributes motor commands. Usually, each nerve contains both motor and sensory axons. Exceptions are the optic and olfactory nerves, which only collect sensory information. The PNS can be further divided into sections called the somatic system and the autonomic system. The somatic system receives information from the external senses (such as touch and vision) and sends motor commands to the skeletal muscles. This system deals with reflexes and voluntary movements.

▶ BRAIN
Human
The human brain consists of the brain stem, the cerebellum, and the cerebrum, which has four lobes. Body activities are controlled by specific areas within the brain.

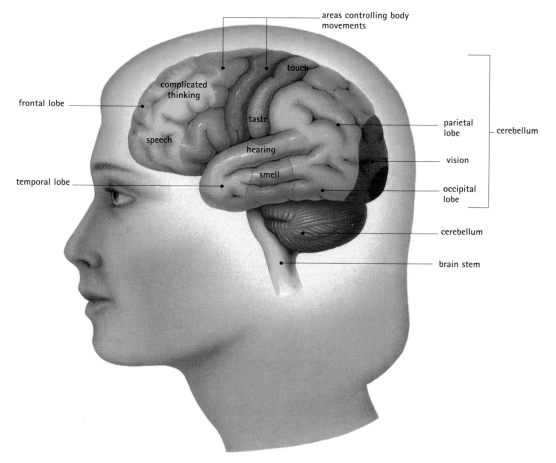

areas controlling body movements

complicated thinking

touch

taste

frontal lobe

speech

hearing

smell

parietal lobe

cerebellum

vision

temporal lobe

occipital lobe

cerebellum

brain stem

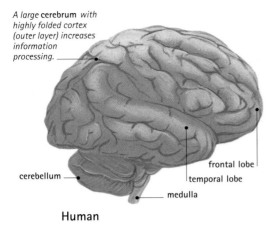

A large cerebrum with highly folded cortex (outer layer) increases information processing.

cerebellum

frontal lobe
temporal lobe
medulla

Human

◀ COMPARING BRAINS

This diagram shows the typical brain structures of a human, a bird, a reptile, and a fish. All vertebrate brains have certain features in common, including a cerebellum to coordinate movement, optic lobes for sight, and olfactory lobes to interpret smells.

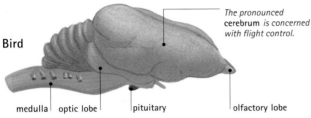

The pronounced cerebrum is concerned with flight control.

Bird

medulla | optic lobe | pituitary | olfactory lobe

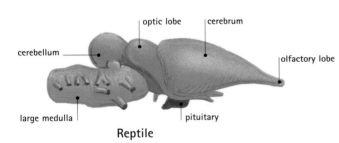

optic lobe | cerebrum

cerebellum

olfactory lobe

large medulla | pituitary

Reptile

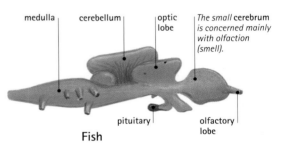

medulla | cerebellum | optic lobe

The small cerebrum is concerned mainly with olfaction (smell).

pituitary | olfactory lobe

Fish

Get rhythm

Some behaviors, such as eating, breathing, and locomotion, are rhythmic. In many of these cases, there is a special network of neurons that produces a basic rhythmic neuronal output. These networks are called central pattern generators. In most cases, this basic rhythm has to be started, stopped, and fine-tuned by ongoing sensory input. For example, a basic locomotor rhythm is generated in the spinal cord of vertebrates and then altered by sensory feedback and input from higher centers in the brain to produce walking or running, depending on the circumstances.

The autonomic system receives information about the internal state of the body and then sends commands to maintain these conditions at appropriate levels. The responses controlled by the autonomic system are usually not under voluntary control. Examples are the size of the pupil of the eye, salivation, heart rate, and control of the digestive and excretory systems. Opposite aspects of each response are controlled by separate parts of the autonomic system. These are the sympathetic and parasympathetic systems. The sympathetic system, for example, controls dilation of the pupil of the eye, and the parasympathetic system controls constriction of the pupil.

The blood–brain barrier

The circulatory system plays a vital role, bringing oxygen and nutrients to the brain. However, the brain, more than any other organ, must also be protected from changes in concentrations of substances and from infections. This is accomplished by a layer of brain endothelial (lining) cells that are connected to each other very tightly. This layer is called the blood–brain barrier. The blood–brain barrier forms a physical barrier to some molecules, bacteria, and viruses. The barrier also regulates which substances can enter or leave the brain. When the blood–brain barrier is breached, as in bacterial meningitis, the results can be very serious or even fatal. However, the blood–brain barrier also stops many medicines from reaching the brain, so drug molecules have to be designed very carefully.

▶ CEREBRAL LOBES
Human

The cerebral cortex of the human brain is divided into four lobes: frontal, temporal, parietal, and occipital. Each lobe has distinct functions, some of which are shown in this diagram.

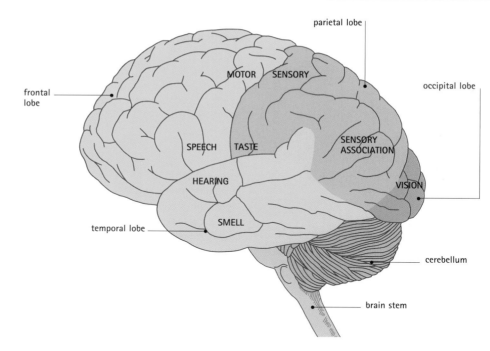

▼ SPINAL CORD
Human

This diagram shows paired spinal nerves leaving the spinal cord. Sensory nerves enter the spinal cord via the dorsal root, and motor nerves exit by way of the ventral root. The sympathetic chain of ganglia runs parallel to the spinal cord.

CNS: The spinal cord

The spinal cord extends down the back of vertebrates and is protected by vertebrae, which are the bones that make up the spine. The major function of the spinal cord is to carry information to and from the brain. The spinal cord can also generate the basic motor rhythms for breathing and locomotion. It also provides some relatively simple responses, such as reflexes. A reflex is an automatic response to a specific stimulus that can occur without sending or receiving any information to the brain. In the knee-jerk reflex, the leg is struck just below the knee. This stretches a tendon and activates sensory neurons that directly synapse with, and activate, motor neurons. These, in turn, cause muscles to contract, jerking the lower leg forward. All this happens without the need for any more complex integration. Most reflexes also involve spinal interneurons between the sensory and motor neurons.

CNS: The hindbrain and midbrain

The rearmost section of the brain is called the hindbrain. That is where the brain and spinal cord connect. Just in front of this junction is a region of the hindbrain called the medulla oblongata, or medulla. The medulla coordinates several autonomic functions, including digestion, breathing, vomiting, and

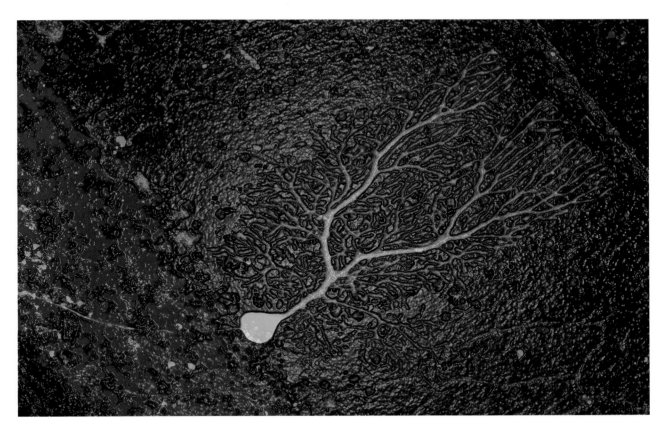

swallowing. The other major section of the hindbrain is the cerebellum (named for the Latin "little brain"). This is a domelike structure behind, and in the case of mammals located beneath, the cerebrum (part of the forebrain). The cerebellum receives sensory information about the position of the body, as well as input from the visual and auditory systems. In addition, higher brain centers in the cerebrum relay information about motor commands to the cerebellum, which uses this information to maintain balance and provide automatic coordination of movements.

In front of the hindbrain is the midbrain. Overall, the midbrain integrates sensory information, especially vision and hearing. Specific areas deal with each of these: the inferior colliculi for hearing and the superior colliculi for vision. In mammals, however, visual information is dealt with mostly by the cerebrum, and the role of the superior colliculi is reduced. The reticular formation regulates the state of arousal in vertebrates. Arousal is the amount of interest in—or

awareness of—the environment. For example, the lowest state of arousal is sleep, and the highest occurs when predators are sensed.

CNS: The forebrain

The most complex neuronal processing is accomplished by the forebrain. This consists of a number of areas associated with different functions. The thalamus receives and sorts many types of sensory information and

▲ *This micrograph shows a Purkinje nerve cell. Large numbers of these highly branched cells occur in the outer layer of the cerebellum. The axons (nerve fibers) of Purkinje cells carry information processed in the cerebellum to the brain stem.*

EVOLUTION

Evolution of the vertebrate nervous system

Increasing size is one major trend in the evolution of the vertebrate brain. Among fish, amphibians, and reptiles, the size of the brain is broadly constant, relative to body size. In birds and mammals, the most recently evolved vertebrates, the brain is comparatively large. A small mammal will have a much bigger brain compared with a reptile of the same size. The cerebrum, in particular, is larger in birds and mammals; in some mammals, the outer layer (cortex) is folded, which increases information processing.

▶ CEREBRAL HEMISPHERES
Human

The cerebrum has two halves, or hemispheres, which process information from opposite parts of the body. For example, when the eyes are fixed on a point straight ahead, information from the left field of vision of each eye is sent to the right cerebral hemisphere, and information from the right field of each eye is sent to the left cerebral hemisphere.

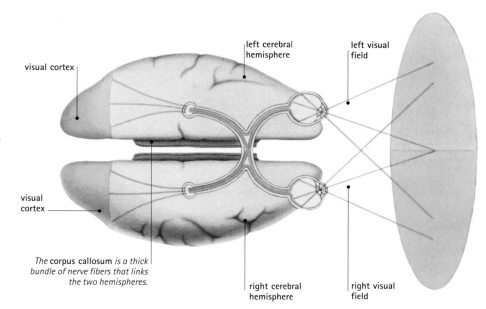

visual cortex

left cerebral hemisphere

left visual field

visual cortex

*The **corpus callosum** is a thick bundle of nerve fibers that links the two hemispheres.*

right cerebral hemisphere

right visual field

Crossed wires

The axons descending from the brain to the spinal cord cross sides as they pass through the hindbrain. As a result of this arrangement, motor commands from the left side of the brain actually go to the right side of the body. The visual system is set up in this way, too, so information from your right eye ends up going to, and being processed by, the left side of your brain. The region where nerve fibers cross hemispheres, and hold the two hemispheres together, is the corpus callosum.

then relays it on to higher brain regions for further processing and interpretation. The hypothalamus is vital in maintaining the internal body state at optimal conditions. It regulates body temperature, hunger, thirst, and sexual functions. The limbic system is a group of structures that influence basic emotional responses. Two parts of the limbic system, the hippocampus and amygdala, are thought to be important in learning and memory.

The frontal part of the vertebrate brain is called the cerebrum. In fish and amphibians, this is relatively small and mostly concerned with processing olfactory (smell) information. In many mammals, the cerebrum is the largest part

◀ SENSORY CORTEX

The surface of the cerebrum, the cortex, has areas that receive incoming sensory information and other areas that are concerned with outgoing motor information. The different parts of the body have different amounts of the cortex devoted to them. The diagram shown is a map of the sensory and motor cortexes, proportional to how much of the cortex is devoted to each body part.

SENSORY CORTEX

lower arm
wrist
hand
fingers
eye
face
teeth, gums, and jaw
pharynx

elbow
head
toes
upper arm
thumb
nose
shoulder
lower lip
tongue
abdomen

genitals
foot
leg
hip
trunk
neck

MOTOR CORTEX

hip
trunk
shoulder
knee
ankle
toes
hip
trunk
hand

fingers
thumb
neck
brow
eyelid and eyeball
face
lips
jaw
tongue
swallowing

of the brain and has many folds. These folds, or convolutions, increase the surface area of the outer layer, which is the cerebral cortex. Since the neuronal cell bodies are all located in the cerebral cortex, a greater surface area allows more neurons and therefore greater processing power.

The cerebrum is the largest and most conspicuous feature of the human brain, with many large convolutions. The human cerebrum consists of left and right cerebral hemispheres, connected by a nerve tract called the corpus callosum. The functions of different regions of the cerebrum have been mapped to some extent using various imaging techniques. Two examples are the motor cortex and sensory cortex. Motor output and sensory information (mainly touch) from different parts of the body can be mapped out across these cortices. This mapping is not uniform, because different body regions vary in their relative importance. For example, the areas in the motor cortex devoted to hands and face are larger than those for the entire torso and legs. Other specific regions of the cerebrum are devoted to processing sensory input from the remaining senses: smell, taste, hearing, and vision. Further regions are active during particular activities, such as reading, speech, and hearing.

Higher brain functions

The higher brain functions most commonly associated with humans, such as memory, emotion, speech, and thought, have been—and continue to be—the subject of extensive scientific investigations. The anatomical centers associated with some of these processes are known, but scientists still have little knowledge of their detailed working.

The different hemispheres of the brain control different functions. For example, the left hemisphere controls speech, language, and calculation, whereas the right hemisphere controls spacial awareness and creative thinking. The processing of separate aspects of complex abilities, such as language and speech, is often distributed around the brain. For example, different sections of the brain are active when someone is listening to a word, thinking of a word, or saying a word. These regions must interact in complicated and fast, yet subtle, ways, even in the most simple conversation between humans.

Imaging living nervous systems

There are several modern ways of imaging the nervous system, in addition to conventional X rays. Positron emission tomography (PET) measures positrons emitted by minute amounts of a radioactive substance administered to a patient. PET scans can be used to measure levels of metabolic activity or blood flow to different areas of the brain. These indicate which parts of the brain are most active at any time. Computed tomography (CT) uses X rays to generate a series of images of "slices" through the head. These can then be built up into a three-dimensional image. An electroencephalogram (EEG) uses electrodes attached to the surface of the head to map the electrical activity in different parts of the brain.

▼ NORMAL BRAIN
This PET scan of a human brain shows a large area (brown) of normal activity.

▼ BRAIN WITH DEMENTIA
In someone with the degenerative brain disorder senile dementia, there is reduced brain activity.

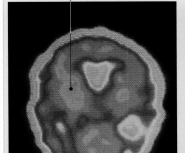

large area of activity

reduced area of activity

The nervous system enables humans to accumulate and use memories and have creative and imaginative thought processes and makes us aware of our surroundings, ourselves, and each other. It also enables humans to communicate complex ideas and strive to understand them. A fuller understanding of these processes is one of most challenging and interesting fields in science today.

RAY PERRINS

FURTHER READING AND RESEARCH
Freeman, Scott. 2001. *Biological Science*. Prentice Hall Publishers: Saddle River, NJ.

Reproductive system

One of the essential characteristics of life is the ability to reproduce. In fact, all organisms that are living today exist because their parents and the parents before them, in generations going back to the first living being, were able to reproduce. The ones that failed to reproduce left no offspring and have no descendants.

DNA and RNA

All the information that is needed to reproduce a complex organism is contained within its DNA (deoxyribonucleic acid). Sections of DNA, called genes, contain code to build individual proteins. This code is copied into a closely related molecule called messenger RNA (ribonucleic acid; mRNA). The mRNA is then "read" by ribosomes, organelles within cells. Ribosomes are "machines" that translate the code in the mRNA to the corresponding sequence of different amino acids—the building blocks of proteins. Proteins, in turn, are functional units that carry out many of the tasks an organism needs to survive, grow, and reproduce.

However, DNA may not have been present at the dawn of life, although RNA may have been. The discovery that RNA molecules can help make other RNA molecules has led to the theory that the first forms of life on Earth may have been formed from self-replicating strands of RNA. By producing copies of themselves, these self-replicating RNA molecules would have become ever more abundant. Accidental errors introduced during the replication resulted in the occurrence of a variety of different RNA molecules, some of which started to carry the necessary information to produce proteins. This "knowledge" made the replication of RNA faster and more efficient. Because of the improved efficiency, RNA molecules that carried this information became much more numerous than the molecules that did not.

With time, an ever-increasing number of proteins were encoded by the RNA molecule, and at each step those RNA molecules that were the most successful in reproducing themselves became the most abundant. As the raw materials to build these systems started to become limited, the less successful forms disappeared. Thus, very early in the history of life on Earth, reproduction may have become a vehicle for evolution; every time a change was introduced to the RNA that enabled it to be replicated more efficiently, this change tended to prevail in the lineages of future generations. At some point in time, RNA started to be copied to DNA, which took over as the storage medium for genetic information. This DNA encoded for proteins that speeded up its reproduction, and it is the ancestor of all life today.

Life-forms that are more effective in reproducing become more numerous, and those that are less successful fade. This process has created an amazing diversity of strategies to reproduce effectively, including many seemingly strange and fascinating features.

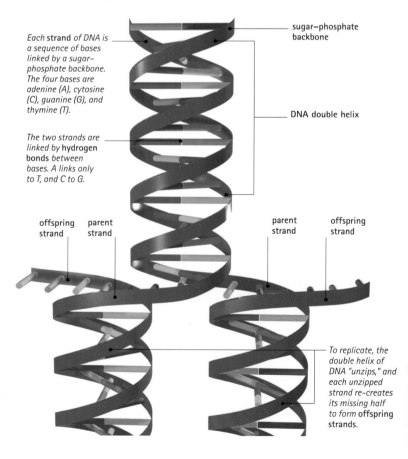

Each **strand** of DNA is a sequence of bases linked by a sugar–phosphate backbone. The four bases are adenine (A), cytosine (C), guanine (G), and thymine (T).

The two strands are linked by **hydrogen bonds** between bases. A links only to T, and C to G.

sugar–phosphate backbone

DNA double helix

offspring strand · parent strand

parent strand · offspring strand

To replicate, the double helix of DNA "unzips," and each unzipped strand re-creates its missing half to form **offspring** strands.

▶ **DEOXYRIBONUCLEIC ACID (DNA)**
All life, except some RNA viruses, uses DNA as the storage medium for genetic information. DNA is a long, threadlike molecule, composed of two intertwined strands. Sections of DNA, called genes, contain the code to build individual proteins.

128

If the driving force behind reproduction is to spread the genetic material of an individual, then the most efficient form of reproduction could be thought to be one where an exact copy, or a clone, of the individual is produced. This form of reproduction, which takes place without mating and in which the offspring is identical to its single parent, is called asexual reproduction. Many species reproduce asexually, and asexual reproduction has several advantages over sexual reproduction, which involves the combination of genetic material from two individuals into the offspring. First, asexual reproduction is the most direct and efficient way to propagate the genes (units of inheritance) of an individual. Second, asexual reproduction requires no mating to produce offspring, saving resources that otherwise would be put into courtship, for example. The third advantage is that there is no need to find a mate. This advantage is probably the reason why asexual reproduction is common among sessile animals (those that cannot move about) and those that live in dispersed populations where the chances of finding a mate are small.

Sexual reproduction

Despite the advantages of asexual reproduction, most multicellular organisms are able to reproduce sexually. During sexual reproduction, sex cells, or gametes, called eggs and sperm are produced by females and males, respectively. Almost all the body cells of most sexually reproducing organisms (exceptions include some plants) are diploid. Diploid cells contain two complete copies of the genome, which is the complete genetic information contained within DNA in a cell's nucleus. One copy of the genome comes from the mother and one from the father.

In contrast, eggs and sperm are haploid, with only a single copy of the genome. In sexual reproduction, two haploid gametes from two individuals fuse to make a diploid zygote, which then grows into an embryo. Sexual reproduction is universal among complex organisms, and this fact suggests that it brings an evolutionary advantage. By mixing genetic material from two parents, sexual

▲ Reproduction is the production of individuals similar to the parent or parents. In sexual reproduction, which occurs in most plants and animals, including cheetahs (above), male and female sex cells fuse to make offspring with genes from both parents.

reproduction produces genetic variation among offspring. The environment is unpredictable and liable to change drastically, causing some individuals to die. Others are likely to survive, because their particular combination of genes makes them able to cope with change. The chance that these offspring will be able to reproduce and pass on their genes is probably far greater than if all the offspring were asexual clones of one parent. To repeat: the fact that sexual reproduction is common clearly indicates that it must bring a huge evolutionary advantage.

FEATURED SYSTEMS

ASEXUAL REPRODUCTION This type of reproduction involves the production of offspring with only one parent. Asexual reproduction may be a part of a complex life cycle that also involves sexual reproduction at some stage. Alternatively, asexual reproduction in a species may continue for many generations or be the only type of reproduction. *See pages 130–133.*

SEXUAL REPRODUCTION During sexual reproduction two sex cells, or gametes, from two individuals fuse to start the development of a new individual that contains genetic material from both parents. *See pages 134–137.*

REPRODUCTIVE STRUCTURES IN ANIMALS Sexual reproduction requires structures and behaviors that enable the sex cells from two individuals to meet. The variety of structures and behaviors in animals and plants is incredibly large. *See pages 138–143.*

REPRODUCTIVE STRUCTURES IN PLANTS Asexual and sexual reproduction in plants requires specialized structures. *See pages 144–147.*

DEVELOPMENT OF THE ZYGOTE After fertilization, the zygote develops into an embryo, which continues developing, often within the mother. *See pages 148–149.*

Asexual reproduction

COMPARE asexual reproduction in a *JELLYFISH* with that in a *SEA ANEMONE*.

COMPARE the regeneration of lost body parts in a *STARFISH* and a *CRAB*.

CONNECTIONS

Asexual reproduction is widespread. Many organisms have complex life cycles, which may involve sexual reproduction at one stage and asexual reproduction at another. Different ways of reproducing asexually include fission, regeneration, and parthenogenesis.

Fission is the division of an organism into two organisms. In binary fission, a single-cell organism, such as a diatom, undergoes cell division to form two almost equal "offspring" cells. More complex forms of fission occur in multicellular organisms. Fission can take place at almost any stage in the life cycle of an organism. Adults can divide by fission through budding or stolonization (the growth of side stems to form new plants). In polyembryony, embryos form asexually in a plant seed or when an animal zygote splits, for example as in identical twins. Larval replication is fission during a larval life stage. It is relatively common among parasitic invertebrates with a complex life cycle.

Many jellyfish and some hydrozoans (such as obelias) alternate between life stages with sexual and asexual reproduction. The sexually reproducing form is the jellyfish stage, also called a medusa. The asexually reproducing larval form of jellyfish is called a polyp, the body shape that is normally associated with hydrozoans, such as *Hydra*. The medusa is the dominant phase in the life cycle of jellyfish, whereas hydrozoans spend most of their life in the polyp form. In both groups, however, the medusa exists as either a female or a male, which mates to form free-swimming larvae, or planulae. The larvae settle on the seabed, where they transform into the polyp form. In jellyfish, the polyp is called a scyphistoma. This repeatedly divides itself transversely, to form a structure called a strobila, which is composed of a stack of saucerlike structures called ephyrae. These break off from the strobila, one at a time, to form young medusae. All the medusae formed from a single

► **Whiptail lizards**
These lizards live in the arid grasslands of the southern United States and northern Mexico. In some populations of this species, there are only females, and young are produced parthenogenically— from unfertilized eggs. To activate egg development, the lizards pair up, undergo a rough courtship, and perform a sexual act in which they place their cloacas (reproductive openings) together. However, since neither animal is male, no sperm are transferred from one to the other.

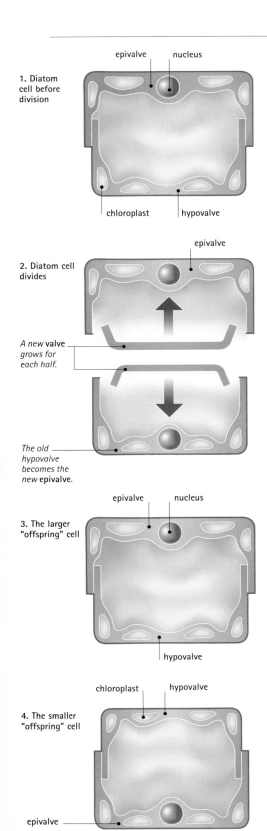

1. Diatom cell before division

epivalve   nucleus

chloroplast   hypovalve

2. Diatom cell divides

epivalve

A new valve grows for each half.

The old hypovalve becomes the new epivalve.

3. The larger "offspring" cell

epivalve   nucleus

hypovalve

4. The smaller "offspring" cell

chloroplast   hypovalve

epivalve

CLOSE-UP

Asexual reproduction of blood flukes

Blood flukes are parasitic flatworms that cause an illness called schistosomiasis, or bilharziasis, in humans living in some tropical regions. The adult blood flukes live in veins of the intestine or bladder, depending on the species of fluke, where they mate and the females lay eggs. These eggs leave the host with the feces or urine. If the eggs end up in water, they hatch as free-swimming larvae miracidia. If a miracidium penetrates a freshwater snail, it becomes a sporocyst. The sporocyst reproduces asexually by fission to produce numerous tadpolelike larvae, called cercariae, which leave the snail. The cercaria are free-swimming, and if they meet human skin they penetrate it. After entering a human, they are carried in the bloodstream to the lungs, the liver, and finally the veins of the intestine or bladder, where they mature sexually and begin a new life cycle.

◀ **Diatom**
A diatom is a single-cell organism that reproduces both asexually and sexually. To reproduce asexually, the cell divides in two to form two "offspring cells." The cells are unequal in size, owing to the nature of the diatom's complex, silica-rich cell wall.

strobila are genetically identical and the same sex, because they are products of asexual reproduction. However, the medusae live independently and can reproduce sexually.

Some hydrozoans produce free-swimming medusae, but the freshwater hydra does not have this stage. Instead, the freshwater hydra reaches adulthood and reproduces sexually in the form of a polyp. However, the freshwater hydra also reproduces asexually. In the hydra, a new individual forms as an outgrowth from the body stalk of an adult individual. Through

▶ **Moon jellyfish**
Asexual reproduction occurs in the polyp stage of the jellyfish life cycle. The polyp, or scyphistoma, divides asexually to form a strobila—a stack of saucerlike structures called ephyrae. One by one, ephyrae break off to become medusae—free-swimming forms of the jellyfish.

IN FOCUS

Polychaete worms: Only the tails have sex

In the Syllidae family of polychaete worms, there are species in which parts of the body break off and engage in sexual reproduction. In the palolo worm, for example, a segment in the middle of the body grows into a second head, and the animal divides into two parts. The front part of the worm regenerates its mid and tail region and continues to live. The newly generated head is in charge of the rear end of the "parent" worm, which swells with sperm or eggs. The rear sections of many palolo worms swim to the surface, where they search for a mate and—if successful—spawn and die. Spawning is risky because the aggregation of worms attracts predators. However, since the front part of the worm remains hidden in sediments on the seabed, the only thing it risks is its tail.

this budding process, a complete hydra is generated. This asexually produced clone eventually breaks off from the parent animal.

Regeneration is the regrowth of body parts from a lost piece of the body. In some cases, this process can result in the generation of two or even several new organisms. If a starfish loses one of its arms, a new one can grow back. Moreover, the amputated arm—unless it was eaten or otherwise completely destroyed—can grow into a complete starfish.

Parthenogenesis, which means "virgin origin," is the process by which an egg develops into an intact individual without first being fertilized by sperm. Parthenogenesis is relatively common throughout the animal kingdom, and it is the only form of asexual reproduction that is

▶ *In some dandelions, seeds are produced asexually. In this type of reproduction, called apomixis, pollination and fertilization do not occur. Seeds develop from ovules containing diploid cells, rather than from the fusion of haploid egg and pollen cells.*

IN FOCUS

Alternation of generations in water fleas

Many species alternate between parthenogenesis (asexual reproduction from unfertilized eggs) and sexual reproduction. For example, in water fleas, parthenogenic females are produced when living conditions are favorable. When conditions worsen, males are produced parthenogenically. These males mate with females, which then produce eggs contained within a thick-walled capsule, or ephippium. When the female loses its exoskeleton, the ephippium is released and sinks to the bed of the lake in which the water fleas live. Ephippia can survive freezing and drying and can lie dormant for more than 100 years. They can even pass unscathed through the gut of a predator. When living conditions improve, the eggs hatch, giving rise to young water fleas.

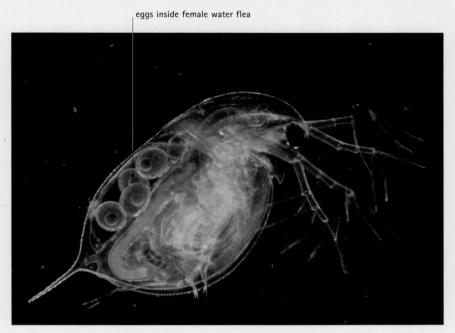

eggs inside female water flea

▲ *When conditions are favorable, water fleas produce female young by parthenogenesis (a form of asexual reproduction). If conditions worsen, males are produced. The males then mate with females, which produce eggs that lie dormant until conditions improve again.*

known to occur in vertebrates. Parthenogenic reproduction sometimes requires sexual activity even if this does not lead to fertilization of the egg. For example, egg development in whiptail lizards, which live in the southern United States, is triggered by courtship. However, some populations of whiptail lizards are made up entirely of parthenogenic females. In order to activate the development of eggs, some females (which have already laid eggs) behave as males: they court the female-acting partner and even engage in mating. As both partners are females, no sperm are exchanged, and the development of the eggs is triggered by the act of mating alone.

Asexual reproduction in plants

In plants, asexual reproduction is common. It is often divided into two types: apomictic and vegetative. Apomictic reproduction involves offspring produced by unfertilized flowers and is comparable to parthenogenesis in animals. In a similar way, vegetative reproduction is comparable to fission in animals and involves the generation of new plants from parts of the plant that are not the germ cells. These include "runners" such as stolons, which are side stems that run from the parent plant above the soil and lay down roots from which new plants arise. Strawberries are a well-known example of a plant that produces stolons. In many plants the runners emanate from the parent plant in the soil, in which case they are referred to as rhizomes. A potato is the enlarged nutrient-filled tip of a rhizome, and humans have known for centuries that planting a potato, which is a stem and not a root, gives rise to new potato plants. In many plants, almost any part of the plant, including the stem and leaves, can generate a new complete individual plant.

Sexual reproduction

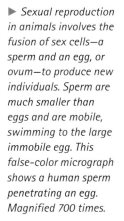
COMPARE sex determination in a *HUMAN* with that in a *HONEYBEE*. In humans, it is a result of sex chromosomes: females have two X chromosomes, and males have an X and a Y chromosome. In honeybees, sex is determined by the number of sets of chromosomes: females have two sets (diploid) and males have just one set (haploid).

Sexual reproduction is based on the presence of females and males within a species. Females produce relatively large and immobile gametes (reproductive cells) called eggs, or ova; males produce sperm, or spermatozoons, which are smaller, mobile, and usually much more numerous.

The DNA in each cell of a eukaryotic organism (all life-forms except bacteria and viruses) is efficiently packed into dense structures called chromosomes. Chromosomes can be divided into autosomes and sex chromosomes. The latter differ between the sexes and contain the genes that lead to sex differentiation. In mammals and many other animals, the female and male sex chromosomes are called X and Y chromosomes, respectively.

Cells can have either a single copy of each chromosome or two almost identical copies. If a cell has only one copy of each chromosome, it is said to be haploid (n). During sexual

Sex determination in various animals		
	male	female
birds	ZZ	ZW
grasshoppers	XO	XX
humans, fruit flies	XY	XX
honeybees	haploid	diploid

reproduction, two gametes—an egg and a sperm cell—fuse to form a zygote, which has a double set of chromosomes, one coming from the egg and the other from the sperm. Cells that have a double set of chromosomes are called diploid (2n).

In many animals and plants, the major stage in the organism's life cycle is made up almost entirely of diploid cells, with only the gametes being haploid. A diploid zygote is created when two gametes fuse. This is called fertilization, and it usually occurs sometime

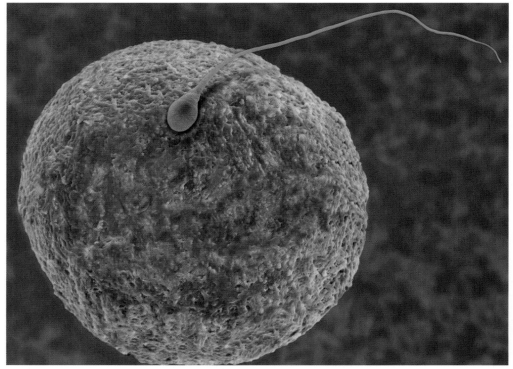

▶ *Sexual reproduction in animals involves the fusion of sex cells—a sperm and an egg, or ovum—to produce new individuals. Sperm are much smaller than eggs and are mobile, swimming to the large immobile egg. This false-color micrograph shows a human sperm penetrating an egg. Magnified 700 times.*

after mating, or copulation. Haploid gametes are formed from diploid cells, through a special kind of cell division called meiosis. Usually when cells divide, the chromosomes replicate and each offspring cell obtains the same chromosome setup as the parent cell. In this way, a diploid parent cell will produce two diploid offspring cells, and a haploid parent cell produces two haploid offspring cells; this type of cell division is called mitosis. In organisms in which the adults are diploid, haploid eggs or sperm are formed through meiosis. Meiosis actually involves two cellular divisions, although the DNA is copied only once. In this process the chromosome number is reduced from diploid to haploid.

Sex determination

In mammals, sex is genetically determined. A setup of two X chromosomes gives rise to a female, whereas the combination of an X and a Y chromosome results in a male. However, not all species follow this pattern. For example, in birds and some fish the presence of two sex chromosomes of the same kind (the ZZ setup) produces a male, and two different kinds of sex chromosomes (the ZW setup) leads to the development of a female.

The sex of an animal can also be determined by factors other than sex chromosomes. One example is environmental sex determination (ESD). In alligators and some other reptiles, the sex of an offspring is determined by the temperature at which the egg develops.

The vast majority of cells in an animal or a plant are differentiated for various specialized functions, such as building a skeleton, or taking up nutrients from the gut or in a root. These cells are called somatic cells. In contrast, the function of germinal cells is to propagate their DNA by the formation of new individuals. In animals, the germinal cells are defined very early in the embryonic development of an individual. They migrate from tissue close to the rear end of the gut, where they first appear, to the location of the developing ovaries (in females) or testes (in males).

In vertebrates, the female and male gonads (ovaries and testes) and sex organs develop from the same structures of the embryo. Early in the development of a vertebrate, there is no visual difference between females and males.

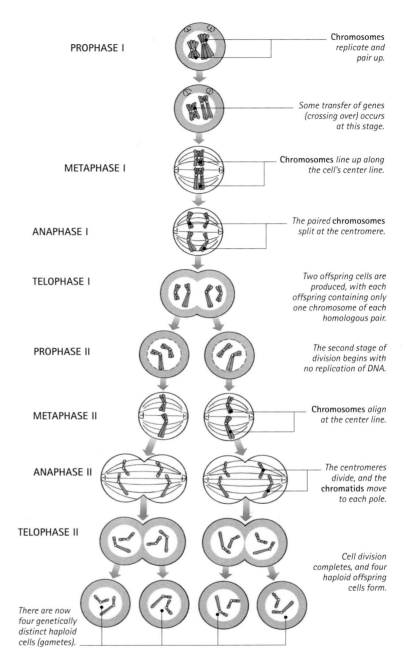

PROPHASE I — Chromosomes *replicate and pair up.*

Some transfer of genes (crossing over) occurs at this stage.

METAPHASE I — Chromosomes *line up along the cell's center line.*

ANAPHASE I — *The paired* chromosomes *split at the centromere.*

TELOPHASE I — *Two offspring cells are produced, with each offspring containing only one chromosome of each homologous pair.*

PROPHASE II — *The second stage of division begins with no replication of DNA.*

METAPHASE II — Chromosomes *align at the center line.*

ANAPHASE II — *The centromeres divide, and the* chromatids *move to each pole.*

TELOPHASE II

Cell division completes, and four haploid offspring cells form.

There are now four genetically distinct haploid cells (gametes).

The development into testes is dependent on a gene on the Y chromosome. Absence of the protein encoded by this gene results in the formation of a female reproductive system.

It is not only the gonads that differ between females and males; there are also differences between the brains of males and females. The development of "brain gender" is important because the brain directs behavior and controls

▲ **MEIOSIS**
Sexual reproduction involves the production of four haploid gametes (sex cells) from one diploid germ cell, in a two-phase process called meiosis.

Male-inducing genes

In humans, the testes-determining factor (TDF) is encoded by the SRY gene, which is located on the Y chromosome. If TDF is present, the undifferentiated embryonic gonads will develop into testes, and if TDF is absent, the gonads will become ovaries. When the testes are formed they start to produce the masculinizing hormone, testosterone, which stimulates the development of male sex organs. Testosterone also depresses the formation of female body parts, including the cells that otherwise would develop into breasts. The SRY gene is central to male development, but there are several other genes that are also believed to contribute to the differentiation between females and males.

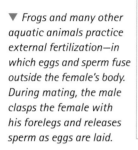

▼ *Frogs and many other aquatic animals practice external fertilization—in which eggs and sperm fuse outside the female's body. During mating, the male clasps the female with his forelegs and releases sperm as eggs are laid.*

functions involved in reproduction, and these behaviors and functions are sometimes very different in males and females. Sex-determining genes probably influence the gender of the brain directly and also indirectly by promoting the secretion of sex hormones.

Egg and sperm development

When the germinal cells arrive at the gonads early in the development of the embryo, they first divide by mitosis. In females, these cells become primary oocytes. The primary oocytes enter the first cell division of meiosis but then stop their development, remaining dormant until they are released from the ovaries during ovulation in the adult female. In humans, the egg does not complete the second cell division of meiosis until it is fertilized by a sperm.

In males, the number of germinal cells, called spermatogonia, in the fetal testes continues to increase by mitosis until birth. The division of diploid spermatogonia then stops until it is

Alternation of generations in plants

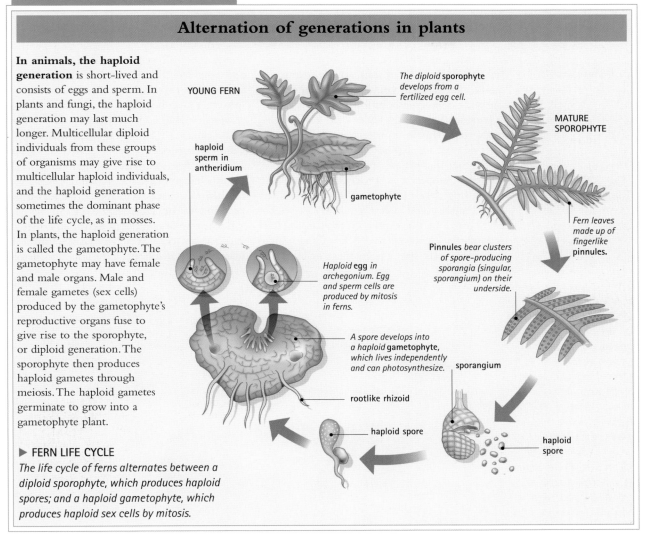

In animals, the haploid generation is short-lived and consists of eggs and sperm. In plants and fungi, the haploid generation may last much longer. Multicellular diploid individuals from these groups of organisms may give rise to multicellular haploid individuals, and the haploid generation is sometimes the dominant phase of the life cycle, as in mosses. In plants, the haploid generation is called the gametophyte. The gametophyte may have female and male organs. Male and female gametes (sex cells) produced by the gametophyte's reproductive organs fuse to give rise to the sporophyte, or diploid generation. The sporophyte then produces haploid gametes through meiosis. The haploid gametes germinate to grow into a gametophyte plant.

▶ **FERN LIFE CYCLE**
The life cycle of ferns alternates between a diploid sporophyte, which produces haploid spores; and a haploid gametophyte, which produces haploid sex cells by mitosis.

YOUNG FERN

haploid sperm in antheridium

gametophyte

The diploid sporophyte develops from a fertilized egg cell.

MATURE SPOROPHYTE

Fern leaves made up of fingerlike pinnules.

Pinnules bear clusters of spore-producing sporangia (singular, sporangium) on their underside.

Haploid egg in archegonium. Egg and sperm cells are produced by mitosis in ferns.

A spore develops into a haploid gametophyte, which lives independently and can photosynthesize.

sporangium

rootlike rhizoid

haploid spore

haploid spore

---

resumed at puberty when the germ cells, now called primary spermatocytes, start to undergo meiosis to produce a constant supply of cells called spermatids, which develop into sperm. Mature sperm are very small and contain, in addition to the haploid genetic material from the male, only what is absolutely required to deliver this material to the egg.

A generalized sperm has a small head, which contains the genetic material. In addition, there is a pouch, or acrosome gland, on the tip of the sperm head, and this is filled with a protein that helps the sperm penetrate the egg. The sperm head is followed by the sperm's midsection, or body, which is packed with mitochondria. These provide the energy required to swim using the sperm's long tail—or in some species several tails. In many water-living species, eggs and sperm are released into the water, and the sperm fertilize the eggs outside the female's body; this is called external fertilization. This type of fertilization is difficult in land-living species partially because the sperm move by swimming. Therefore, in land-living animals fertilization typically occurs inside the female's body where body fluids can provide the medium for sperm and egg to meet. This type of fertilization is called internal fertilization.

Reproductive structures in animals

CONNECTIONS

COMPARE the claspers of a male *HAMMERHEAD SHARK* with the penis of a *DOLPHIN* and the spermatophore of *LOBSTER*.

Compare the brood pouch of a male *SEA HORSE* with the uterus of a female *COELACANTH*.

The sex organs, or testes, of most male mammals (except elephants, bats, and aquatic mammals) are located in a skin pouch, or scrotum outside the body cavity. Sperm probably develop best at a temperature slightly lower than that of the body core. Muscles in the scrotum are able to lift or lower the testes to vary their distance from the body and therefore maintain a constant temperature. Production of sperm (spermatogenesis) takes place in long, highly coiled tubes, called seminiferous tubules, which are located in the center of each testis. The seminiferous tubules have thick walls in which the sperm cells develop. In these walls, there are also Sertoli cells, which provide support and nutrients for the developing sperm cells. Sertoli cells also produce chemical substances, including hormones, that control the formation of sperm and sexual development in the body as a whole. The most immature diploid germ cells (spermatogonia) are located along the edges of the seminiferous tubules, and the more differentiated spermatids are nearer to the hollow center of the tubule with their tails pointing toward the center.

Between the seminiferous tubules there is another cell type, which produces hormones (chemicals that influence the function of cells

▶ MALE REPRODUCTIVE SYSTEM AND FORMATION OF SPERM
Human
Spermatogenesis, or sperm production, occurs in the seminiferous tubules of the testes. Diploid germinal cells divide by meiosis to produce haploid, tailed sperm. Sertoli cells assist spermatogenesis by providing support and nutrients. Sperm are ejaculated through the urethra during sexual intercourse.

▶ SPERMATAZOON
Human
Mitrochondria in the sperm's tail produce energy for the lashing flagellum, which propels the sperm. The acrosome at the front of the head allows the sperm to penetrate an egg.

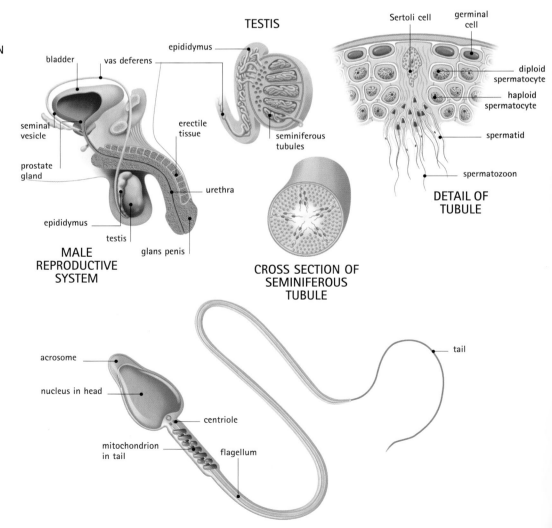

TESTIS

bladder · vas deferens · epididymus

seminal vesicle · prostate gland · epididymus · testis · glans penis · erectile tissue · urethra · seminiferous tubules

MALE REPRODUCTIVE SYSTEM

CROSS SECTION OF SEMINIFEROUS TUBULE

Sertoli cell · germinal cell · diploid spermatocyte · haploid spermatocyte · spermatid · spermatozoon

DETAIL OF TUBULE

acrosome · nucleus in head · mitochondrion in tail · centriole · flagellum · tail

elsewhere in the body). These cells are called Leydig cells, and they make the masculinizing hormone, testosterone, which stimulates the Sertoli cells inside the seminiferous tubules to direct the development of sperm.

When the sperm cells are almost mature they are released into the central lumen of the seminiferous tubules. They are moved along the seminiferous tubules to a storage area, or epididymis, which is located next to the testis. There, the sperm cells go through their final maturation and are now ready to be mixed with other components in a fluid called semen.

Sperm and semen are expelled from the body via tubes called the vas deferens and the urethra, the latter ending at the tip of the penis. Along the way, other fluid secretions are added by the bulbourethral glands, the seminal vesicles, and the prostate gland. To allow sperm to enter the female's reproductive tract, the male's penis is inserted into the vagina of the female.

Female mammalian reproductive system
The ovaries of a female mammal are located in the posterior (rear) part of the body cavity. In the ovaries, the developing egg is surrounded by supportive cells. Some of these cells, called theca cells and granulosa cells, produce and secrete feminizing sex hormones. Collectively, the egg and its supporting cells make up a follicle.

At ovulation, eggs are released from the ovaries into the funnel-like opening of the oviduct (the fallopian tubule). Hairlike cilia lining the inner walls of the oviduct sweep the eggs toward the uterus. If sperm are present in the female's reproductive tract, they will usually intercept the egg in the upper part of the oviduct. To get there the sperm will have traveled from the upper parts of the vagina, where they were

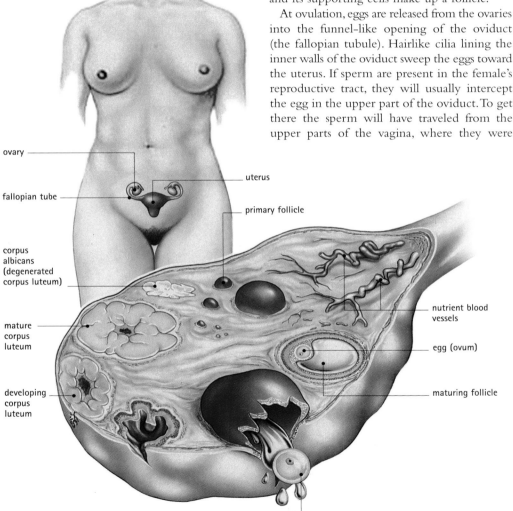

ovary

fallopian tube

uterus

primary follicle

corpus albicans (degenerated corpus luteum)

mature corpus luteum

developing corpus luteum

nutrient blood vessels

egg (ovum)

maturing follicle

mature egg (ovulation)

◄ EGG DEVELOPMENT IN HUMANS
Human females have two ovaries, each containing follicles, in which eggs develop. When an egg is released (ovulation), the empty follicle develops into a corpus luteum. This secretes the homone progesterone, which causes the wall of the uterus to thicken in preparation for implantation of a fertilized egg. If this does not occur, the corpus luteum degenerates.

139

ejaculated from the male, through the cervix, which separates the vagina from the uterus, and along the uterus into the oviduct.

In humans, the opening of the vagina (and the opening of the urethra through which urine passes) is framed by two pairs of skin folds: the labia minor and labia major. The labia minor are thin and richly supplied with sensory nerves that respond strongly to touch. Where the two labia minor meet in front of the urethra is a knob of erectile tissue, called the clitoris, which has the same origin in the embryo as the penis in males. Around the labia minor are the labia major,

which have thicker skin and protect the delicate tissues situated underneath. In human infants the opening to the vagina is partly covered by a thin membrane called the hymen. The hymen often ruptures or stretches in childhood, for example during physical exercise; or this may happen during the first sexual intercourse.

Vertebrate sex hormones

Hormones are chemicals that are produced in one part of the body and are carried in the bloodstream to other parts of the body where they have an effect. In vertebrates, the release

▼ **Female emperor dragonfly**
Eggs develop in the two ovaries and pass along the oviduct to the bursa. Dragonflies practice internal fertilization. The male has a structure called an aedeagus, which is used to introduce sperm into the female's bursa. In females, sperm is stored in spermathecae and released to fertilize eggs.

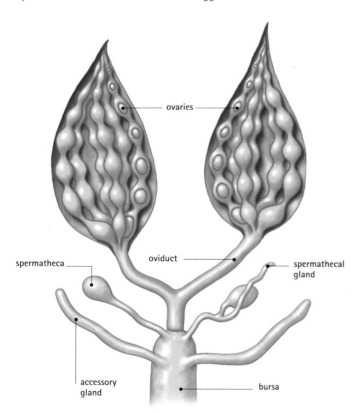

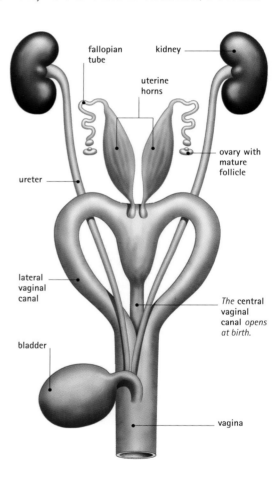

▲ **Female kangaroo**
Like many other female mammals, kangaroos have two ovaries in which eggs develop. However, kangaroos are unique in having two wombs, or uterine horns, and two curved vaginas. When a kangaroo is ready to give birth, a new opening develops, through which the joey passes.

of hormones that direct sexual reproduction is coordinated by an area of the brain called the hypothalamus. The hypothalamus receives information from other parts of the brain and produces signals that decide the timing of sexual maturation and ovulation, or the release of eggs from the ovaries.

The hypothalamus releases a hormone called gonadotropic-releasing hormone (GnRH). This hormone acts on a small gland, the pituitary gland, or hypophysis, which lies just below the hypothalamus. The pituitary gland produces several hormones, and two of these, follicle-

IN FOCUS

Timing of reproduction

In many animals, sexual reproduction occurs only during a limited period of time every year, and it is therefore important that females and males are prepared to reproduce at the same time. Day length and temperature are common cues for sexual maturation in animals that live in nontropical regions. In the tropics, there may be other triggers, such as the phase of the moon, drought, rain, or flood. Chemical signals (pheromones) released by individuals that are ready to breed are also very important in coordinating sexual reproduction within a population.

▼ Male hammerhead shark
Sharks have a single opening, called a cloaca, through which they excrete wastes and reproduce. Sperm produced in the testes travels along Wolffian ducts to the cloaca. The male transfers sperm into the female using one of his claspers, which has a groove along which the sperm travel.

▼ Male Jackson's chameleon
All lizards and snakes have two penises, called hemipenises, one on either side of the cloaca. During mating, sperm pass down the spermatic ducts to one of the hemipenises, which is inserted into the female's cloaca. Fertilization of eggs by sperm takes place in the female's oviducts.

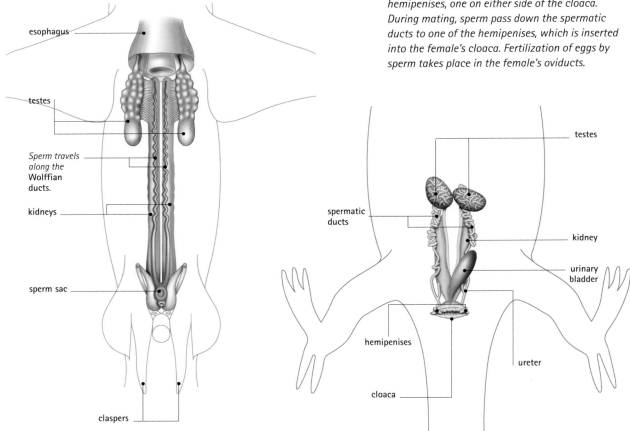

esophagus

testes

Sperm travels along the Wolffian ducts.

kidneys

sperm sac

claspers

spermatic ducts

testes

kidney

urinary bladder

hemipenises

ureter

cloaca

Hermaphroditism

Hermaphroditism is the state of an individual having both male and female sex organs. There are two kinds of hermaphroditism: simultaneous, in which the organism has male and female organs at the same time; and sequential, in which the animal changes from male to female, or vice versa. Some coral reef fish are hermaphrodites. Clown fish, for example, live closely associated with sea anemones. Only the biggest clown fish associated with a suitable sea anemone is female. She breeds with the second

biggest fish, which is the only adult male. If the female disappears, the sexually mature male will change sex and become female; one of the previously immature fish will mature and become the new breeding male.

In some fish species a single breeding male has a harem of females, and each individual starts its adult life as a female and may later change sex to become a male. The queen angelfish follows this strategy. It would make little sense to be a small male queen angelfish because the dominant male does not tolerate

other males in his territory. Therefore, most of the other fish in the group are females until the dominant male disappears. When this occurs, the largest of the females changes sex and becomes a male.

▼ *Clown fish can change sex. If the largest fish in a group (the breeding female) dies, the only breeding male then changes sex to become the new female. One of the immature fish becomes the new male.*

stimulating hormone (FSH) and luteinizing hormone (LH), respond to GnRH. In males, LH stimulates the production of testosterone in the Leydig cells of the testes. Testosterone stimulates the growth of the testes and promotes the bodily changes associated with male puberty. These changes vary greatly among different species and include an increase in muscle mass

and hair growth, changes in coloration, and, in humans, a deepening of the voice. FSH acts with testosterone on the Sertoli cells inside the seminiferous tubules to stimulate the production and maturation of sperm cells.

In female mammals, FSH and LH stimulate production in the ovaries of several female sex hormones, collectively called estrogens, which

IN FOCUS

Testosterone and birdsong

Every breeding season, male birds sing to claim their territory, compete with other males for dominance, and attract females. As the breeding season approaches, and before the males start to sing, the testes begin to expand. The growing testes produce more and more testosterone. The testosterone stimulates further testis development and sperm maturation and produces changes in the brain. For example, testosterone stimulates the growth of several parts of the brain required for singing.

► *Lincoln's sparrow is widespread in North America. The male bird has a rich, gurgling song that it uses to attract a mate.*

in turn stimulate the maturation of the eggs and the development of female sex characteristics throughout the body.

In mammals, these female sex characteristics include the growth of the mammary glands (breasts) and, in humans, widening of the hips and distribution of fat tissue to the buttocks and thighs. In vertebrates that lay eggs or otherwise have embryos that are not nourished from the mother through a placenta, estrogens are also responsible for stimulating the production of a yolk-protein precursor in the liver. This protein is then transported in the bloodstream to the ovaries, where it is incorporated into the developing egg and converted to yolk.

Another key function of estrogen hormones in female mammals is to stimulate the growth of the inside lining of the uterus. This lining is called the endometrium. It is within the endometrium that the embryo embeds and develops into a fetus.

Corpus luteum and progesterone

As the egg or eggs (depending on the species) mature and the follicle grows, the production of estrogens increases. At a certain point in time (12 days after the start of menstruation in humans), the pituitary gland produces a surge of the hormones LH and FSH. This surge leads to the ovulation of a mature egg or eggs from the ovaries. The ruptured follicle is left behind in the ovary, and it transforms into a structure called a corpus luteum. The corpus luteum continues to produce estrogen and a hormone called progesterone. The endometrium depends on estrogens and progesterone for continued growth and maintenance, in preparation for the implantation of a fertilized egg.

Shedding of the endometrium

If the egg or eggs do not become fertilized, the corpus luteum withers away (this happens within two weeks after ovulation in humans). As a result of the drop in levels of estrogen and progesterone that follows, the endometrium starts to break down.

In most species of mammals, the endometrium is absorbed by the body, but in certain primates, such as humans, chimpanzees, and gorillas, the thickened lining of the uterus is shed through the vagina, a process called menstruation. If an egg is successfully fertilized by sperm, growth and development of the embryo continue in the endometrium.

Reproductive structures in plants

Plant reproduction differs fundamentally from animal reproduction because plants have a multicellular haploid phase of the life cycle as well as a multicellular diploid phase. In some groups of plants, such as mosses and liverworts (bryophytes), the haploid stage, which is called the gametophyte because it bears gamete-producing sex organs, dominates the life cycle; and in other groups (notably flowering plants, or angiosperms) the diploid stage, which is called the sporophyte because it produces spores, dominates.

Gymnosperms were the first seed-bearing plants to evolve. There are perhaps fewer than 750 gymnosperm species alive, but some of these are very widespread. The most familiar gymnosperms are conifer trees, which include pines and spruces. The sporophytic scale tissue of the seed cone and the gametophyte inside constitute the ovule. The term *gymnosperm*, meaning "naked seed," refers to the ovule of the plants in this group, which is not protected by an ovary or fruit tissue as it is in flowering plants, or angiosperms. *Angiosperm* means "enclosed seed."

Flowering plants

Angiosperms are the most diverse group of living plants, with more than 257,000 species. Angiosperms have flowers for at least part of the year, and these are the plants' sexual organs. A typical flower consists of sepals, petals, male stamens, and a female carpel. The carpel contains the ovules and seeds. Stamens typically consist of a thin stalk, or filament, with an anther at the end. The anther contains

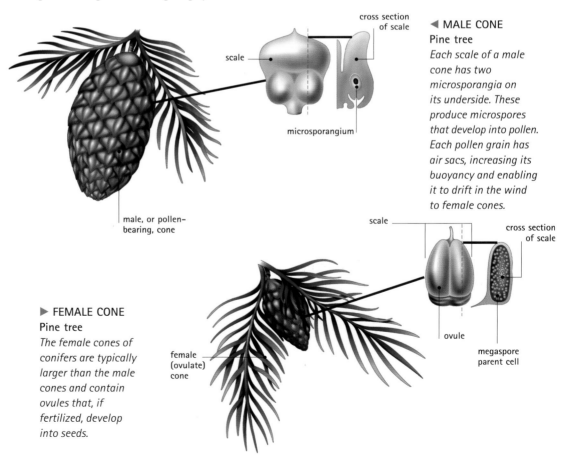

cross section
of scale

scale

microsporangium

◄ MALE CONE
Pine tree
*Each scale of a male
cone has two
microsporangia on
its underside. These
produce microspores
that develop into pollen.
Each pollen grain has
air sacs, increasing its
buoyancy and enabling
it to drift in the wind
to female cones.*

male, or pollen-
bearing, cone

scale

cross section
of scale

► FEMALE CONE
Pine tree
*The female cones of
conifers are typically
larger than the male
cones and contain
ovules that, if
fertilized, develop
into seeds.*

female
(ovulate)
cone

ovule

megaspore
parent cell

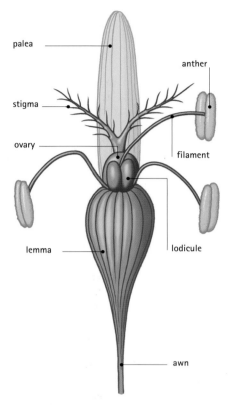

palea

anther

stigma

ovary

filament

lemma

lodicule

awn

▲ **Marsh grass flower**
Grass flowers are wind-pollinated and do not need showy petals to attract insects. Two scales, the palea and lemma, initially enclose the male and female organs. The scales are split apart by the swelling of two small bodies called lodicules.

IN FOCUS

Gamete production in flowering plants

In the anthers of a flower, diploid cells called microsporocytes divide meiotically to form haploid microspores, which in turn develop into pollen grains. Pollen grains have two nuclei: a generative nucleus and a tube nucleus. (After pollination, the generative nucleus divides mitotically to make two sperm nuclei.)

Within the ovule of the carpel, a diploid megaspore parent cell divides through meiosis to produce four haploid megaspores. Each megaspore divides mitotically four times to produce eight haploid nuclei, all enclosed within one single cell, called a megagametocyte. Two groups of three nuclei each migrate to opposite ends of the megagametocyte and two nuclei stay in the middle. In this rearrangement of nuclei, cell walls are built around the six nuclei that are located at the ends of the megagametocyte. One of these cells becomes the egg (which, if pollinated, fuses with a sperm nucleus from the pollen). The two nuclei in the center of the megametocyte become enclosed within the same cell wall to produce a single cell containing two polar nuclei (which, if pollinated, fuses with the other sperm nucleus from the pollen).

the pollen-producing microsporangia, or pollen sacs. Megasporangia are housed in the carpel. In a single flower, there may be one or more carpels, which may be fused. The carpel has a stigma where pollen may attach. The stigma is often at the end of a style, which connects the stigma to the ovary at the base of the whole structure. The ovary, style, and stigma are together called the carpel.

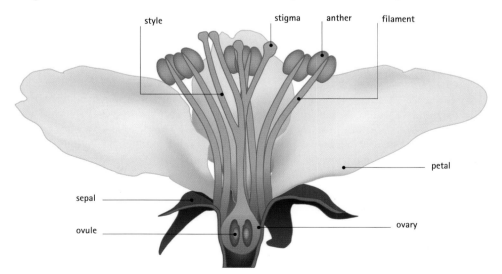

style

stigma   anther   filament

petal

sepal

ovule

ovary

◀ **Apple flower**
Flowers with both male and female reproductive organs, such as apple, are called "perfect." The female parts are called carpels and are situated in the center of the flower. Each carpel is made up of a stigma, style, and ovary. The male parts, which surround the carpels, are called stamens; each one is composed of an anther and filament.

▲ *Many flowers are pollinated by bats. Such flowers open, or remain open, during the night and emit strong, fruity smells to attract bats. Above, a long-tongued bat, which occurs in North and South America, visits a cup-and-saucer plant.*

IN FOCUS

Fertilization in flowering plants

When a pollen grain lands on the stigma of a compatible carpel, it develops a pollen tube and burrows into the stigma and down the style until it encounters the ovule. Chemical signals—probably released by cells in the ovule—serve as a guide for the pollen tube. These chemicals lead the pollen tube to a small opening in the ovule. At first, the tube contains two haploid nuclei. One is the generative nucleus and the other is the tube nucleus. The haploid generative nucleus divides by mitosis to produce two haploid sperm nuclei. Both sperm nuclei are released into the ovule; one of the sperm nuclei fuses with the egg to produce the zygote of the new diploid sporophyte generation. Curiously, the other sperm nucleus fuses with the central cell in the ovule that already contains two nuclei. The fertilized central cell now contains three sets of chromosomes; it is triploid (3n). The triploid cell rapidly divides through mitosis and produces a tissue that serves as nutrient for the developing plant embryo. The process is called double fertilization and is unique to angiosperms.

However, there is considerable variation in this generalized flower structure. For example, many angiosperms have female-only or male-only flowers, lacking carpels or stamens, respectively. Flowers that have both carpels and stamens are called "perfect," and those that lack either male or female structures are called "imperfect." If an individual plant has both female and male flowers, it is called dioecious. Otherwise, plants are either female or male and are called monoecious. In either case, it is the sporophyte generation that produces flowers. The gametophyte of flowering plants is highly reduced—the female gametophyte consists of only seven cells.

Unlike other plant groups, such as mosses and ferns, angiosperms and gymnosperms are not dependent on water for the male gamete to travel to the egg. Instead, the male gametes of angiosperms and gymnosperms sit in pollen grains that may be carried by wind or by animals (for example, insects, bats, and birds) that make contact with the flowers.

Following fertilization, there is a program of cell division and development in the embryo and in the tissues that surround it. The cell layers surrounding the ovule develop into a protective seed coat. The embryo itself develops into a stage in which buds develop that will form the seed leaves. Later, the seed will lose most of its water content (up to 95 percent), and it stops growing and becomes dormant. The embryo remains in this quiet state until conditions are favorable for it to germinate.

IN FOCUS

Reproduction in a gymnosperm

Pine trees provide a good example of gymnosperm reproduction. Pine trees have two types of cones: seed cones and smaller strobili. Seed cones produce megaspores, which are the female gametophytes; and the strobili produce microspores, which are male gametophytes. Microspores develop into pollen grains, which travel on the wind to sticky seed cones. The pollen grain then germinates to form a pollen tube, which makes its way—over several months—to the female gametophyte. One of the pollen cells then divides to produce two sperm cells, one of which fuses with the egg cell to form a zygote. Each scale of the seed cone may contain several eggs, which when fertilized can produce several zygotes. In most cases, only one of these survives. Each zygote develops into an embryo within a seed. Pine seeds are wing-shaped, so they can be carried by the wind away from the parent plant to grow in new location.

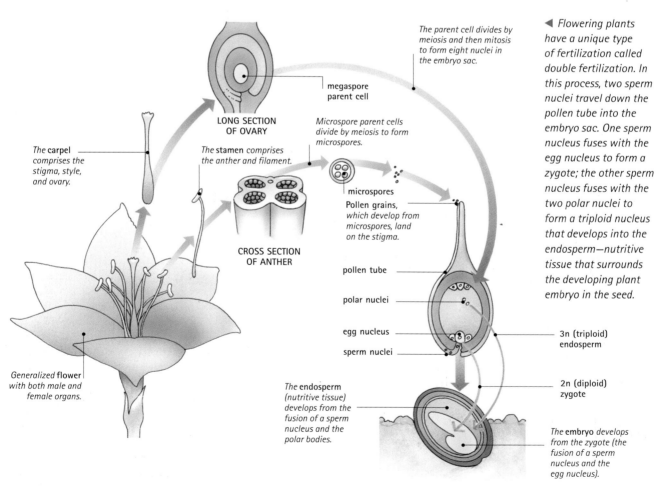

The parent cell divides by meiosis and then mitosis to form eight nuclei in the embryo sac.

megaspore parent cell

LONG SECTION OF OVARY

Microspore parent cells divide by meiosis to form microspores.

The carpel comprises the stigma, style, and ovary.

The stamen comprises the anther and filament.

microspores

CROSS SECTION OF ANTHER

Pollen grains, which develop from microspores, land on the stigma.

pollen tube

polar nuclei

egg nucleus

sperm nuclei

Generalized flower with both male and female organs.

The endosperm (nutritive tissue) develops from the fusion of a sperm nucleus and the polar bodies.

◀ Flowering plants have a unique type of fertilization called double fertilization. In this process, two sperm nuclei travel down the pollen tube into the embryo sac. One sperm nucleus fuses with the egg nucleus to form a zygote; the other sperm nucleus fuses with the two polar nuclei to form a triploid nucleus that develops into the endosperm—nutritive tissue that surrounds the developing plant embryo in the seed.

3n (triploid) endosperm

2n (diploid) zygote

The embryo develops from the zygote (the fusion of a sperm nucleus and the egg nucleus).

Development of the zygote

In animals, a sperm penetrates an egg using proteins stored in the acrosome to make a hole. The sperm then loses its tail, and the membrane that surrounds the egg cell nucleus breaks down, allowing the DNA of the two gametes to combine. The resulting zygote divides repeatedly without gaining size until a species-specific number of cells have been formed. This initial bout of cell divisions results in the formation of a ball of cells called a blastula. The blastula has a fluid-filled cavity, called the blastocoel, around which much of the embryo development occurs. The formation of a blastula marks the completion of the first major step in the development of an embryo.

The second milestone in the developmental process is gastrulation. During gastrulation, the wall of the embryo forms a pouch that reaches deep into the blastocoel. The opening to this pouch is the pore that in many animals, including all vertebrates, becomes the anus. The cavity inside the pouch will become the intestine, and the wall of the pouch (the endoderm) will also give rise to other organs, such as the liver and lungs. From this primitive gut, two pockets appear and pinch off, creating a hollow sac on either side of the gut. The two sacs are the coelomic vesicles, and the cells that make up the sacs (the mesoderm) will form

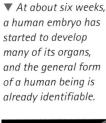

▼ *At about six weeks, a human embryo has started to develop many of its organs, and the general form of a human being is already identifiable.*

▲ *Female dogfish release egg sacs—commonly called mermaid's purses—in which embryos develop for several months before hatching. The embryos are nourished by a large reserve of yolk.*

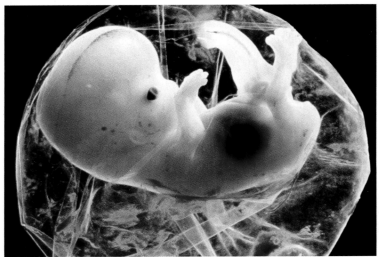

internal tissues such as muscles, blood vessels, reproductive organs, and the inside lining of the body cavity. At this stage, the ball of cells is called a gastrula. The outer cell layer of the gastrula is called the ectoderm, and it will give rise to the body surface and superficial structures such as hair and nails, the nervous system, and some other tissues.

In vertebrates, the third major stage of zygote development is the formation of a nervous system. First, the notocord, which develops into the backbone, is laid down along the ceiling of the gastrula. The nervous system is formed from the ectoderm (outer cell layer) overlaying the future backbone. Above the rod of backbone cells, the ectoderm thickens and

GENETICS

Hox genes

The differentiation of the body along the longitudinal axis is at least partially determined by a large family of genes, which in vertebrates are called Hox genes. The proteins that are encoded by these genes are present at different concentrations and in different combinations along the longitudinal body axis. This unequal distribution of Hox gene products decides where the brain, spinal cord, front limbs, hind limbs, and other body parts will form.

flattens to form a neural plate. Ridges develop along the longitudinal edges of the neural plate, and these ridges rise and fold toward the middle, where they fuse, generating a neural tube. The neural tube develops into the central nervous system. The front end of the neural tube becomes the brain, and the rest becomes the spinal cord.

During the formation of the neural tube, cells from the endoderm grow into longitudinally repeated segments on either side of the neural tube. These segments of endodermal tissue are called somites. Early in the development of the embryo, the somites are more or less uniform, but they become successively more diverse. The somites later give rise to several of the structures that are unevenly distributed along the body, including the ribs, vertebrae, limbs, and muscle. Once the task of system and organ development has been accomplished, the remaining period of development is primarily characterized by growth. In humans, the final phase is called the fetal period.

Gestation and the placenta

In animals that bear live young, such as most mammals and some fish and reptiles, the period from fertilization of the egg until birth is called gestation. In mammals, the length of gestation is generally related the animal's life span. For example, in short-lived mammals, such as small rodents, gestation may last less than a month, whereas in long-lived species, such as elephants, it can be as long as 22 months. In humans, gestation is called pregnancy and takes about 40 weeks (9 months). The length of gestation is also related to the size of the fetal skull that will fit through the mother's pelvis during birth.

All mammals except monotremes (such as the platypus) and marsupials (such as kangaroos and the koala bear) are placental. The placenta is the organ by which the developing embryo attaches to the wall of the uterus. The placenta develops from the chorion (the outermost layer of cells of the blastula). It is attached to the lining of the uterus and connects to the growing fetus by the umbilical cord. Oxygen, nutrients, and antibodies pass across from the mother's blood—via a thin layer of cells—into the fetus's blood; and fetal waste products pass into the mother's blood for excretion by her kidneys and lungs. No direct mixing of fetal and maternal blood occurs. In this way, the fetus is nourished until birth. The placenta is then expelled in the afterbirth.

CHRISTER HOGSTRAND

▼ **COMPARISON OF EMBRYOS**
Amphibian, bird, and mammalian embryos divide, or cleave, at different rates and have structural differences right from the outset of development. For example, bird embryos quickly develop a

Amphibian   Bird   Mammal

polar bodies
AT FERTILIZATION
yolk mass
zona pellucida
FIRST CELL DIVISION
blastodisk (yolk-free region)
yolk
BLASTULA (EXTERIOR)
blastoderm
blastocoel
yolk
blastocyst cavity
inner cell mass
subgeminal space
BLASTULA (VERTICAL SECTION)

FURTHER READING AND RESEARCH
Marshall Graves, Jenny. 2004. *Sex, Genes, and Chromosomes.* Cambridge University Press: Cambridge, UK.
Purves, W. K., G. H. Orians, D. Sadava, and H. C. Heller. 2003. *Life: The Science of Biology*, 7th ed. W. H. Freeman and Company: New York.

Respiratory system

Energy is a requirement for life. Organisms get the energy they need through a process called respiration, which generally requires the presence of an essential gas—oxygen. Cells need oxygen to react with molecules from food, typically in the form of the sugar glucose. The reaction liberates energy; it also leads to the formation of molecules of water and another gas, carbon dioxide. This gas must be removed from the organism; the release of carbon dioxide and the simultaneous uptake of oxygen are called gas exchange. The liberation of energy using oxygen is called aerobic respiration. Without it, few of an organism's life processes could take place.

An anaerobic life

However, cells in tissues such as muscle sometimes use up oxygen faster than it can be supplied—for example, during intense exercise. Cells then switch to a different method of energy production called anaerobic respiration. Again, glucose is broken down to release energy, but with a chemical called lactic acid created as a by-product. Anaerobic respiration cannot continue for long; lactic acid builds up and must be removed. This buildup causes symptoms such as a stitch and muscle fatigue. However, anaerobic respiration can be crucial for providing a sudden burst of speed. This ability is so important that large sections of an animal's body can be devoted to providing for these occasional anaerobic bursts. Fish often have small bundles of red aerobic muscle running along their body. This is used for slow swimming. Vast banks of white anaerobic muscle flank the red muscle; they kick into action when the fish pursues prey or avoids a predator.

Why bother with oxygen?

Anaerobic respiration is also important for many microorganisms. Some, such as yeast, release alcohol as a by-product. Other microorganisms that respire anaerobically (anaerobes) produce lactic acid as a by-product, and the bacteria that live in the intestines of ruminants, such as sheep, cattle, and deer, generate methane. So, if energy can be generated anaerobically, why do organisms bother with oxygen and the challenge of gas exchange at all? This is because aerobic respiration is a far more efficient process; it produces 19 times more energy for each molecule of glucose.

How is oxygen produced?

Like animals, fungi, and most other organisms, plants and algae also liberate energy through aerobic respiration, in which they take in oxygen and release carbon dioxide

▼ GAS EXCHANGE IN DIFFERENT ORGANISMS

Gas exchange occurs in a variety of ways: diffusion directly across the cell surface or skin; across structures called papulae in animals such as starfish; along the tracheal system of insects; or across gills in fish or alveoli in the lungs of mammals.

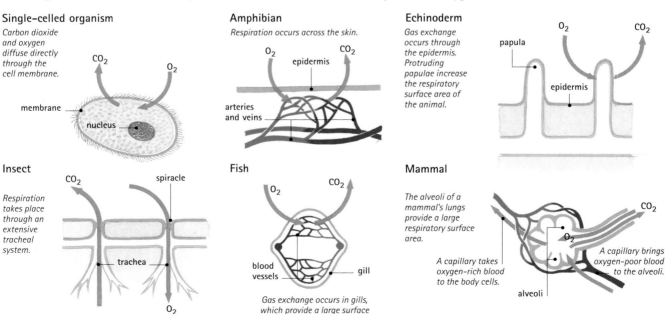

Single-celled organism
Carbon dioxide and oxygen diffuse directly through the cell membrane.

CO_2 O_2 membrane nucleus

Amphibian
Respiration occurs across the skin.

O_2 epidermis CO_2 arteries and veins

Echinoderm
Gas exchange occurs through the epidermis. Protruding papulae increase the respiratory surface area of the animal.

papula O_2 CO_2 epidermis

Insect
Respiration takes place through an extensive tracheal system.

CO_2 spiracle trachea O_2

Fish
O_2 CO_2 blood vessels gill

Gas exchange occurs in gills, which provide a large surface area for respiration.

Mammal
The alveoli of a mammal's lungs provide a large respiratory surface area.

CO_2 O_2 *A capillary takes oxygen-rich blood to the body cells.* *A capillary brings oxygen-poor blood to the alveoli.* alveoli

▲ *The elephant seal holds the record as the deepest-diving marine mammal, reaching depths of 5,180 feet (1,580 m) and routinely holding its breath for over an hour without ill effects.*

through pores called stomata. However, the movement of gases in these organisms changes over the course of a day. During daylight hours, plants take in carbon dioxide and release oxygen. They fix carbon dioxide and water to form sugars. This process, called photosynthesis, is driven by the energy of sunlight. The sugars are later stored or respired.

Oxygen is a by-product of photosynthesis and is released. Without photosynthesis, atmospheric oxygen would not occur and multicellular life as we know it could not exist.

What are the requirements for gas exchange?

Gas exchange depends on a process called diffusion. In this process, molecules move along a diffusion gradient—from a point of high concentration to a point of low concentration. Thus, for oxygen to diffuse across a respiratory surface, its concentration must be lower on the inside of a respiratory surface than on the outside.

Efficient gas exchange has some other requirements. The surface across which exchange takes place must be thin, to minimize the distance gases need to diffuse. It must also be moist: gases dissolve into the fluid. A moist respiratory surface is not a problem for aquatic organisms, because they are surrounded by water; but the need for moisture and the accompanying problems of water loss are important for terrestrial organisms. The final requirement is a large surface area; this requirement more than any other has shaped the evolution of respiratory structures.

Breathing through the skin

Microorganisms exchange gases through diffusion across their outer membrane. As oxygen is used up, more diffuses in; when carbon dioxide levels rise, some carbon dioxide diffuses out. Such diffusion across the outer membrane—

FEATURED SYSTEMS

BREATHING UNDERWATER Most water-living, or aquatic, animals breathe using gills—thin-walled respiratory surfaces across which gases diffuse into and from the water. *See pages 154–156.*

BREATHING AIR USING LUNGS Most four-legged animals depend on lungs for gas exchange. Lungs contain moist, thin-walled membranes and have a vast surface area across which gases diffuse. *See pages 157–158.*

THE TRACHEAL SYSTEM Insects, centipedes and millipedes, and many spiders use tracheae, tubes that take air through pores in the exoskeleton to the tissues where it is needed. *See pages 159–160.*

RESPIRATORY PIGMENTS These molecules occur in the blood of many animals and help carry gases around the animal's body. *See pages 161–163.*

VENTILATION Air or water in an animal's respiratory system can quickly become depleted of oxygen, so a continual flow is needed to bring in fresh supplies. *See pages 164–166.*

MAXIMIZING EFFICIENCY Animals and plants improve the efficiency of gas exchange by maximizing uptake rates, increasing surface areas, and avoiding water loss. *See pages 167–169.*

also called cutaneous gas exchange—is sufficient for some animals, especially those with most or all of their cells in contact with the outside, such as jellyfish and sponges. Many insect larvae (immature forms) breathe solely through their surface, as do many aquatic worms and sea slugs.

Changing shapes

For most larger animals, though, gas exchange across the skin cannot provide the body with enough oxygen. This is because the larger an animal is, the smaller its ratio of surface area to volume. The ratio is high for small animals; in other words, there is a lot of skin per unit of volume. However, for larger animals, the ratio gets progressively

▼ A whale's two nostrils are located on the top or back of the head and are more commonly called blowholes (toothed whales such as killer whales have a single blowhole). As the whale surfaces, the blowholes open and air is exhaled explosively through them, forming a misty vapor called the blow. Fresh air is then inhaled, and the blowholes close again.

COMPARATIVE ANATOMY

Giant skin breathers

Although cutaneous (across skin) gas exchange is generally ineffective for larger animals such as vertebrates, there are a few exceptions. Hibernating frogs draw oxygen through their skin as they doze on the bottom of ponds. Frogs use very little oxygen during this time, and diffusion across the skin can provide more than enough. Some salamanders and other amphibians called caecilians rely wholly on cutaneous gas exchange; these unusual amphibians do not possess lungs. However, they can live only in waters that are extremely rich in oxygen. The caecilian *Atretochoana eiselti* is by far the largest lungless terrestrial vertebrate, measuring up to 31 inches (80 cm) long. Not much is known about this amphibian, but it is believed to live in the cool, fast-flowing streams of western Brazil.

lower. The amount of skin available for gas exchange per unit volume decreases until a critical size is reached beyond which skin diffusion is too inefficient.

Larger animals need more efficient methods of oxygen uptake in which there is a greater surface area available for respiration. For example, starfish have projections over their surface called papules to increase surface area. Most aquatic animals, though, exchange gases through respiratory structures called gills. Gills can be internal, such as those of fish or mollusks. Other animals, such as young amphibians and mayflies, have external gills.

Gas exchange on land

Animals with gills rely on oxygen dissolved in the water. This reliance can be problematic: many aquatic insects, for example, are restricted to cool, fast-flowing streams that are high in oxygen. Land animals have access to far richer oxygen resources. However, respiratory surfaces need to be moist, so those of land animals are internal and can generally be shut off from the outside to minimize water loss.

The oxygen story

Oxygen is an extremely reactive gas. Oxides, chemicals formed by the reaction of oxygen with other elements, are all around us. Carbon dioxide, rust (iron oxide), and water (dihydrogen oxide) are examples. That there is any free oxygen in the atmosphere at all is due solely to the output of photosynthesis. Billions of years ago, the atmosphere contained little or no oxygen. The first organisms on Earth were anaerobic—they did not use oxygen to generate energy. Then, around 3.5 billion years ago, a group of bacteria called cyanobacteria appeared. These were the first photosynthesizers, and they had a dramatic impact. Atmospheric oxygen levels rose sharply. The oxygen caused an ecological disaster—almost all of the anaerobes were killed off. They survive today only in extreme habitats such as pond sludge or on hydrothermal vents. The composition of the air we breathe today is a legacy of this ancient apocalypse. Oxygen levels continued to rise, and life evolved to cope with the dramatic change in atmospheric composition. With oxygen (in the form of ozone) forming a layer in the upper atmosphere that cut out harmful ultraviolet (UV) light, more complex organisms began to evolve.

There are two main breathing strategies adopted by land animals. Almost all four-legged animals breathe through lungs. These are organs where air is alternately drawn in and expelled. Oxygen diffuses into blood carried by a network of capillaries. There, it binds to hemoglobin, an iron-containing protein in red blood cells. Hemoglobin also helps remove carbon dioxide, which is released at the lungs. Hemoglobin is a type of respiratory pigment. There are several others in the animal kingdom. Mollusks and crustaceans, for example, have a copper-containing pigment called hemocyanin that gives their blood a blue tint.

An alternative strategy

Insects have a completely different system. Air enters the body through a system of pores, called spiracles, in the exoskeleton. It passes through a series of increasingly small tubes called tracheae. The tips of the finest tubes are blind-ended and moist; oxygen diffuses from there into the tissues that need it. Therefore, almost all insects have no need for respiratory pigments. Myriapods such as centipedes also have tracheae. So do some spiders, although most also have capillary-rich folded structures called book lungs, with respiratory pigments to carry away the oxygen.

Breathing underwater

A quatic animals need to take in dissolved oxygen that is in the water around them. Cutaneous gas exchange can provide enough oxygen for most tiny organisms. However, larger organisms must increase their surface area relative to their volume for this to prove efficient. Cutaneous gas exchange sometimes provides enough oxygen for larger animals with a very low metabolic rate (the rate of energy use by the body), such as hibernating frogs. Turtles, too, use cutaneous respiration to remain underwater for long periods. They pump water into the throat and rectum to take in more oxygen. However, most aquatic animals use respiratory structures called gills.

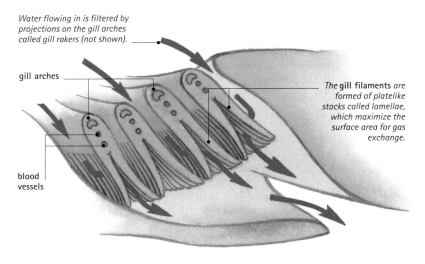

Water flowing in is filtered by projections on the gill arches called gill rakers (not shown).

gill arches

The gill filaments are formed of platelike stacks called lamellae, which maximize the surface area for gas exchange.

blood vessels

What are gills?

Gills consist of a stack of fine branches or folded flaps with a very thin outer membrane. The vastly increased surface area that gills provide allows rapid gas exchange to occur between the water and the inside of the gill's membrane. In most aquatic organisms, blood passes through the gills to take the oxygen away, often with the help of respiratory pigments.

Gills occur in various animal groups, from crustaceans and mollusks to vertebrates such as amphibians and fish. Some animals have internal gills. These can be folded structures such as fish gills, banks of filaments as in bivalves such as the giant clam, or branching trees as in sea cucumbers. Gills may also be external, as in polychaete worms and most aquatic insect

▲ GILLS
Trout
Each gill is made up of numerous gill filaments. The surface of each gill filament is folded into platelike structures called lamellae, creating a large surface area for gas exchange. Each lamella is supplied with blood vessels, which pick up oxygen from the moving water and release carbon dioxide.

After passing over the gills, water is expelled through the exhalant siphon.

In the gills, oxygen passes through the thin gill walls into the blood, and carbon dioxide moves in the other direction.

Water is drawn in through the inhalant syphon.

◄ GILLS
Giant clam
The giant clam draws water into its inhalant siphon and over its gills. There, oxygen passes from the water through the thin walls of the gills and into a series of thin tubes that carry blood around the clam's body.

Breathing atmospheric air

Most aquatic four-legged animals must come to the surface frequently to breathe. The need to visit the surface may be a hindrance for these animals, but for many smaller organisms the surface provides a vital resource. It allows some to live in foul, low-oxygen waters where gills would be hopelessly ineffective. Rat-tailed maggots live in the sludge at the bottom of puddles. They breathe through a long, telescopic siphon that extends from their rear end. Some aquatic fly and beetle larvae in low-oxygen environments breathe by connecting their spiracles to airways inside the stems of submerged plants. Water scorpions live in cleaner waters, but they also breathe through a siphon. The siphon allows the insect to remain still as it waits to ambush prey.

▼ **Rat-tailed maggot**
The larva has an extensible tube, or siphon, through which it draws air into its body. The siphon enables the maggot to live in water that has a very low concentration of oxygen. Rat-tailed maggots feed on decaying plant material.

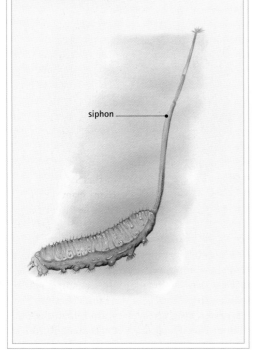

siphon

larvae. Insect gills connect directly to the tracheal system of air tubes. The gills of stonefly nymphs run along the body; those of young mayflies form tufts on the abdomen; and those of damselflies extend from the rectum.

Opposing streams

Water can be saturated with dissolved oxygen, but even then there is much less oxygen available than in the air above the surface. Oxygen levels drop further as the water temperature increases. So gills need to be superefficient to remove as much oxygen from the water as possible. Fish and many other gilled animals maximize uptake by using a countercurrent system. Water flows in one direction over the gills as the animal swims, for example, or by the production of a ventilating current. The blood inside the gill flows in the opposite direction. This ensures that blood takes in oxygen throughout its passage through the gills.

Other ways of getting oxygen

Gills are very common in the animal world, but they do not provide the only solution to getting oxygen underwater. Adult insects do not have gills. Many underwater insects must breathe air, but some store air so that they can minimize the time spent in the hazardous surface waters. These insects have hydrophobic (water-repelling) hairs that trap bubbles of air and hold them against the body. Backswimmers trap air bubbles using long hairs on their

Gill form and function

Although gills evolved to facilitate efficient gas exchange, some animals have additional uses for their gills. Many use gills to help them swim. Young horseshoe crabs use their gills as paddles. Dragonfly nymphs have gills inside their rectum. These gills are ventilated by drawing in water; by squeezing the water out rapidly, the nymph can jet-propel itself forward. Gills may also be used for feeding. Bivalves such as mussels and clams use their gills (or ctenidia) to filter small particles from the water. The particles are sorted by appendages near the mouth called labial palps before being passed into the mouth. Some gills are not used for gas exchange at all. Mayfly nymphs use theirs to beat a current of water over a patch of folded cuticle on their abdomen; gas exchange takes place there instead.

Underwater arachnids

Spiders are an amazingly diverse group of invertebrates. However, very few spiders have evolved for a life in water. A few hunt on the surface film of ponds and pools, but just one species spends its life underwater—the water spider. Lacking gills or paddles, it may seem ill-suited for an aquatic life, but it has an amazing method for getting the oxygen it needs to respire. The spider weaves a sheet of silk and attaches it to underwater plants; this forms a domed "diving bell." The spider's abdomen is covered by hairs that trap air when the animal visits the surface. The spider uses its legs to flick air bubbles into the diving bell.

The water spider rests in its diving bell during the day. At night it sallies out to catch aquatic insects and small fish, beating its legs to drive itself through the water. A thin layer of air around its body acts as a physical gill. Prey is taken back to the diving bell and eaten.

▶ SPINNING A WEB
A water spider spins an underwater sheetlike web between the fronds of an aquatic plant. The spider then surfaces to collect air on its hairy abdomen.

▶ FILLING WITH AIR
The spider uses its legs to flick off the air on its abdomen into the web to form a nest with a bell-shaped bubble. The spider rests in the nest during the day, but hunts at night, eating prey such as tadpoles always in the nest.

underside, as well as under the wings and on the upper surface of the forewings, where their spiracles (airholes) are located. Bubbles can last a long time before they run out of oxygen. This is because as the insect uses up oxygen, more diffuses in from the surrounding water. Such air bubbles are called physical gills.

Some insects, such as *Aphelocheirus* bugs, can remain submerged indefinitely. These bugs have a very dense layer of short, bent hydrophobic hairs coating much of the body. The hairs hold an extremely thin layer of air close to the body. This is called a plastron. It acts as a gill, drawing oxygen from the surrounding water.

Breathing air using lungs

Gills do not work effectively on land. The gill filaments clump together, reducing the surface area available for gas exchange. All land vertebrates, instead, breathe through organs called lungs. These organs contain chambers that provide a thin, moist surface of vast area ideal for gas exchange. Amphibians such as frogs have large but short lungs with a single chamber, although the walls are partitioned. In reptiles, partitioning is more elaborate, and some reptiles, such as snakes, get by with just one lung. The most complex lungs occur in the vertebrates that have the highest demand for oxygen, namely the mammals and birds.

The passage of air

Air normally enters mammals through the nostrils. This allows breathing and feeding to occur simultaneously, although the linkage

How snails breathe

Lungs do not occur solely in vertebrates—some invertebrates have them, too. Land snails evolved from gilled marine ancestors, but they have lost their gills entirely. Instead, their mantle cavity—a space in the snail—has evolved into a lung. Its walls are richly supplied with blood vessels and are strongly ridged to increase the surface area. Some land snails have secondarily become aquatic, living in freshwater. Most must visit the surface regularly to fill their lung, although species that live in oxygen-rich waters, such as pond snails, fill the lung with water and rely on the diffusion of dissolved oxygen across its walls.

with the mouth usually remains. Breathing through the nostrils also allows the nasal hairs to filter the air and structures called nasal turbinates to warm and moisten it.

A flap of tissue called the glottis tilts up to allow air into the larynx, which contains the vocal cords. In swallowing, the glottis tilts down, so food passes into the esophagus

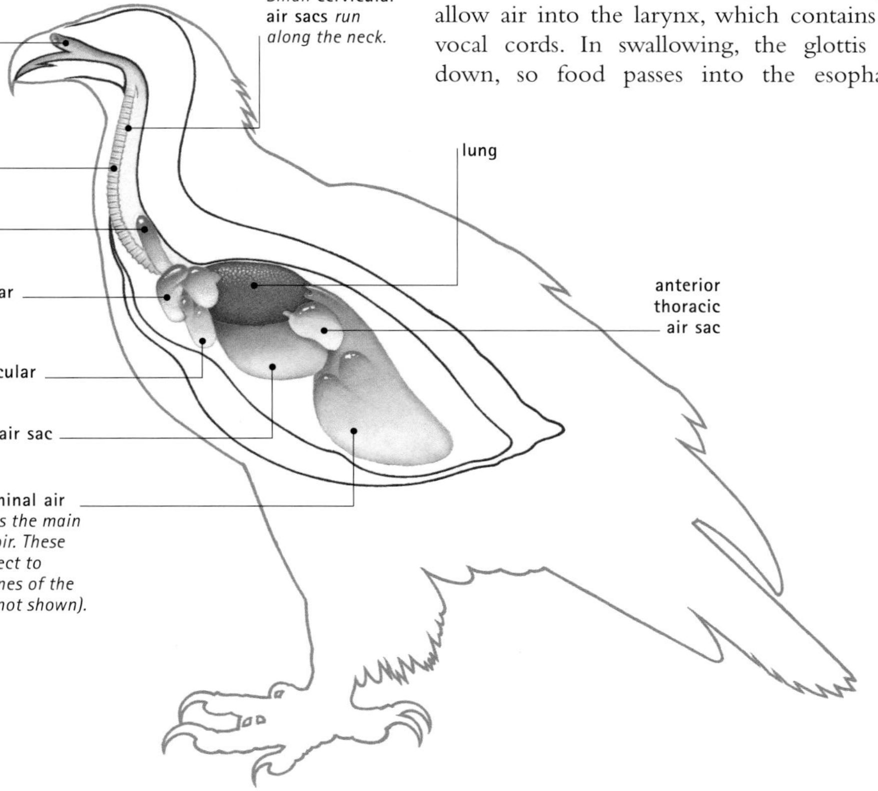

nasal cavity

Small cervicular air sacs *run along the neck.*

trachea

axillary air sac

subscapular air sac

interclavicular air sac

posterior air sac

lung

anterior thoracic air sac

The abdominal air sacs *act as the main air reservoir. These sacs connect to hollow bones of the skeleton (not shown).*

◄ LUNGS AND AIR SACS

Bald eagle
Birds have lungs and also a complex series of air sacs. The lungs and air sacs, along with hollow bones linked to air sacs, allow a continuous stream of air to pass through the lungs in a one-way flow.

Fish with lungs

Tetrapods, or four-legged vertebrates, descended from lobe-finned fish that emerged from the sea about 350 million years ago. These ancient fish must have been able to breathe atmospheric air. There are a few species of fish that can do this today. Some characins can use their swim bladders as a lung, and *Hoplosternum* catfish use part of their gut. The best-known air-breathing fish are the lungfish. Their lungs are derived from the swim bladder and may be similar to those of the earliest tetrapods. Lungfish lungs are divided into pockets richly supplied by blood vessels. Some lungfish have just a single lung; it helps them get enough oxygen in times of shortage so they can survive in oxygen-poor waters. Others have two lungs, and these fish hole up in mud cocoons when their lakes dry up, breathing occasionally through their lungs until the rains return.

▼ *Lungfish lungs evolved from the swim bladder. The fish use their lungs for breathing in seasonal droughts. There are three types of lungfish: South American, African, and Australian. The African and South American species have two lungs, and the Australian (below) has just one lung.*

instead. Air passes through the larynx into the trachea. The trachea is made of smooth muscle reinforced by rings of cartilage. Mucus-producing cells line the trachea; the mucus traps particles, and banks of tiny filaments called cilia carry the mucus up to the throat. At its lower end, the trachea divides into two main airways, each supplying a different lung. These are called the bronchi.

The lungs are in the chest cavity, where they are protected by the rib cage. The bronchus of each lung divides into thousands of smaller tubes called bronchioles. These lead to tiny air sacs, or alveoli. Richly supplied by capillaries, the thin-walled alveoli are the site of gas exchange. The human lung contains millions of alveoli; they give a total surface area for respiration of about 800 square feet (75 sq m).

How is breathing controlled?

The movement of air into and out of the lungs is involuntary, or automatic. The breathing control center is in the medulla, part of the brain stem. Neurons (nerve cells) send signals to the rib and diaphragm muscles to regulate the rate and depth of breathing.

Breathing is affected by a number of factors. Receptors in the aorta (the body's main artery) and carotid (neck) arteries monitor amounts of oxygen and carbon dioxide in the blood and also its acidity. The medulla itself measures levels of carbon dioxide in the fluid surrounding the brain and spinal cord. Receptors that respond to mechanical stimuli (mechanoreceptors) in the lungs monitor how stretched the tissues are, and this prevents the lungs from overinflating. Receptors in the lungs and airways transmit messages to the medulla if irritating particles or chemicals are present. This triggers a sudden exhalation in the form of a cough or sneeze.

Super lungs

Flight is the most energetically demanding form of locomotion. Birds' lungs are very different structurally from those of mammals, and they are also far more efficient. A bird's trachea divides into two bronchi as in mammals, but inside the lung each bronchus separates into two main airways linked by smaller channels called parabronchi. Gas exchange takes place through the walls of the parabronchi, which have a unique structure. They are permeated by fine, branching tubes called air capillaries, along which gases diffuse. The tubes are intertwined with blood vessels.

Connected to the lungs are several air sacs, which fill along with the lungs when the bird breathes in. These are also linked to hollow cavities inside some of the bones. Air sacs act to store fresh air; the air moves into the lungs when the bird breathes out. In this way, birds' lungs can absorb oxygen throughout their respiratory cycle.

The tracheal system

Lungs provide one solution to the challenge of exchanging gases on land. An alternative solution is to have a system of air tubes that carries oxygen directly to the tissues; this is called a tracheal system. Several invertebrate groups have a tracheal system, including the velvet worms, myriapods, most spiders, and the largest animal group of all, the insects. The insect tracheal system is supremely efficient; it needs to be, because insect flight muscle uses oxygen faster than any other known tissue.

How the tracheal system works

Paired pores called spiracles are the gatekeepers of the tracheal system; oxygen and carbon dioxide move in and out through them. Insects such as fleas, bees, and moth larvae have eight or more pairs running along the body; other insects, such as some fly larvae, have just one functional pair, or even, in the case of some aquatic species, none at all. Spiracles are usually kept closed to reduce water loss, and they often contain a hair-fringed plate to keep out dirt.

Breathing inside other animals

Many insects develop inside other animals, and some of these parasites go on to kill their hosts. How do these insects get the air they need? Chalcid wasp larvae, which are internal parasites, or endoparasites, of other arthropods, connect to the air outside through a tube that formed part of their egg. The tube penetrates the host's body wall. Some tachinid fly larvae force their arthropod hosts to grow a structure called a respiratory funnel. This is an envelope of tissue that surrounds the larva and plugs into the host's spiracles. Many endoparasites breathe atmospheric air directly. For example, human botflies burrow into flesh but keep an airway to the outside open through the skin.

Each spiracle opens into a chamber called an atrium, from which the tracheae arise. The tracheae are large, cuticle-lined tubes that carry gases between the spiracles and tissues. Spirals of thick cuticle called taenidia strengthen the tracheae and keep them from collapsing if the

◀ The tracheal system of insects and other land-living arthropods becomes less efficient with increasing size, and therefore limits the size of the animal. The goliath beetle from Africa is the heaviest known insect. It weighs up to about 3 oz (85 g) —which is more than some birds. However, in prehistoric times, some insects, such as dragonflies, were much larger because there was more oxygen in the air then.

The constraints of tracheae

For small organisms, tracheae are vastly superior to lungs. Tracheae deliver oxygen efficiently, and energy is not squandered on powerful pumping hearts or respiratory pigments. However, tracheal systems become less efficient with increasing size, placing a constraint on the maximum size of insects; that is why there are no rhino-size beetles or bird-hunting dragonflies.

Reliance on the tracheal system is also responsible for insects' inability to conquer marine habitats. Just one group lives at sea, the sea skaters, and its members only glide on the surface. The incompressibility of the tracheal system renders aquatic insects buoyant; a marine insect would not be able to dive quickly enough to escape a predator.

▼ **BOOK LUNGS**
Most spiders and scorpions have respiratory organs called book lungs. Oxygen from the air moves across the platelike lamellae into the bloodlike hemolymph; waste carbon dioxide moves the other way.

pressure drops. In certain parts of the tracheae, the taenidia are absent, and the tracheae expand to form collapsible air sacs. These sacs allow ventilation of the system.

Tracheae branch into increasingly narrow airways. Eventually they become fine, blind-ended tubes called tracheoles. Every body cell is close to or in contact with a tracheole. Oxygen diffuses through the thin tracheole membrane

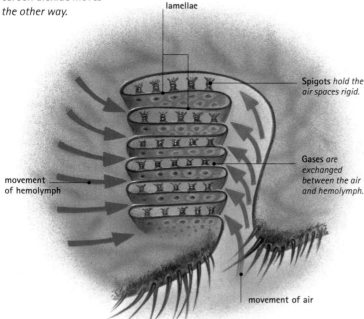

lamellae

Spigots *hold the air spaces rigid.*

Gases *are exchanged between the air and hemolymph.*

movement of hemolymph

movement of air

and dissolves into the fluid surrounding the cells. Oxygen then enters the cells, with carbon dioxide moving in the opposite direction.

Where are these tubes?

Tracheae from neighboring spiracles link to form a pair of major trunks that run the length of each side of the body. Smaller tubes, called tracheoles, branch off to supply the organs and tissues, and they also link the main trunks together. Tracheoles are most abundant in regions of high oxygen demand, such as the wing muscles. During flight, the thorax tracheae are cut off from the rest of the respiratory system. This isolation stops the oxygen requirements of the flight muscles from depriving other essential tissues, such as the brain.

Underwater tracheae

Aquatic insects use tracheae, too. The spiracles of insects that breach the surface to get air, such as backswimmers, are surrounded by hairs that repel water strongly. The spiracles of mosquito larvae bear valves that close when the animal dives. Often only the spiracles near the end of the body are functional in young aquatic insects that visit the surface, although more become active as the insect approaches its final molt. The gills of young caddis flies and damselflies are richly supplied by tracheoles, while the spiracles of some water beetles and bugs open directly into bubbles or films of air on the body.

Spider breathing

Most spiders also have tracheae, although they are less efficient than insect tracheae and work in a different way. They do not deliver oxygen directly to cells; instead, they release oxygen into the bloodlike hemolymph, where it binds to a respiratory pigment. Spider tracheae evolved from structures called book lungs. Book lungs occur in the abdomen. A small atrium enlarges to form a series of horizontal air pockets that are interwoven between stacks of thin-walled leaflets (lamellae) through which hemolymph flows.

Mygalomorphs (tarantulas and relatives) have two pairs of book lungs; in most other spiders, the hind pair has evolved into tracheae, whereas a few groups of spiders have lost their book lungs altogether. The tracheae allow large volumes of air to be stored, and also limit the need for a powerful circulatory system.

Respiratory pigments

In most animals, oxygen binds at the respiratory surface to a molecule called a respiratory pigment. Respiratory pigments link readily to oxygen where it is freely available: in the lungs, for example. This binding property is called oxygen affinity. The binding is reversible, allowing oxygen to be released wherever it is needed for respiration.

Amazing hemoglobin

Hemoglobin is the most widespread respiratory pigment in the animal kingdom, occurring in almost all vertebrates and in invertebrates such as earthworms and echinoderms. It is also by far the most efficient. A hemoglobin molecule consists of four long protein chains, or globins, each coiled into a kidney shape. In the middle of each globin is a molecule with an ion (minute particle) of iron at its center. This molecule is called a heme group. The arrangement of globins permits four oxygen molecules to be carried at any one time. Hemoglobin allows 50 times more oxygen to be transported around the body than could be carried by the plasma (liquid part of the blood) alone.

Path of hemoglobin

The oxygen affinity of hemoglobin is highest in the lungs. There, it is cool and plenty of free oxygen is readily available. The hemoglobin binds tightly to the oxygen and changes from purple red to bright red in the process. How does hemoglobin release oxygen to the tissues that need it? Much depends on the pH, or relative acidity, which is determined by the amount of tiny hydrogen ions present.

As a cell respires, it produces carbon dioxide. An enzyme called carbonic anhydrase converts much of the carbon dioxide into carbonic acid. This acid dissolves into the plasma, forming

COMPARE the respiratory pigment hemoglobin in vertebrates, such as the NEWT, PUMA, and TORTOISE, with hemoglobin in the invertebrate EARTHWORM. In vertebrates, hemoglobin occurs inside red blood cells, whereas in the earthworm, the hemoglobin is free in the blood.

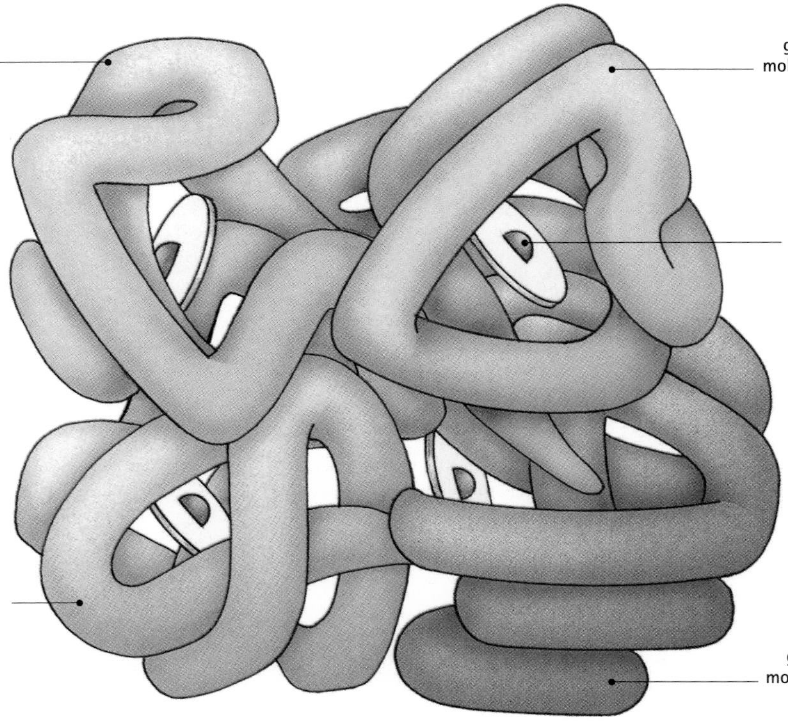

CONNECTIONS

Each globin is a long chain.

globin molecule

globin molecule

globin molecule

heme group

◄ HEMOGLOBIN
Hemoglobin is the respiratory pigment of vertebrates and some invertebrates, including earthworms and starfish. Each hemoglobin molecule is made up of four globin molecules. Each globin has an iron-containing heme group, which transports oxygen and gives blood its red color.

Other vertebrate pigments

Hemoglobin is not the only vertebrate respiratory pigment. A similar pigment, called myoglobin, occurs in muscles. Myoglobin serves as an oxygen store. It binds extremely tightly to oxygen, releasing its cargo when oxygen shortages occur during exercise. Babies in the uterus have a different type of hemoglobin from their mothers. Called fetal hemoglobin, it has a very high oxygen affinity; this allows it to take oxygen from the mother's blood at the placenta.

▼ Red blood cells, or erythrocytes, contain the respiratory pigment hemoglobin. In humans and other mammals, the cells are disk-shape and lack a cell nucleus. In other vertebrates, the cells are oval and have a nucleus (x 2,000).

bicarbonate and hydrogen ions. The hydrogen ions increase the acidity. An arriving hemoglobin molecule meets a warm, oxygen-poor, and—crucially—acidic environment. This combination of factors causes the molecule's oxygen affinity to decrease. The hemoglobin's essential cargo of oxygen is released, diffusing into the cell, and respiration takes place. The influence of the acidity caused by carbon dioxide on hemoglobin is called the Bohr effect.

Helping remove carbon dioxide

The principal role of hemoglobin is oxygen transport, but it plays an important part in the removal of carbon dioxide, too. There are several ways of getting carbon dioxide to the lungs. A small amount of the gas dissolves directly into

Life without pigments

Not all animals need respiratory pigments. Insects have tracheae that deliver oxygen directly to cells, so they do not need pigments. However, there are a few exceptions. Backswimmers are aquatic bugs that use hemoglobin as an oxygen store, releasing the gas to help them maintain buoyancy in the water. Just one group of vertebrates does without hemoglobin: the icefish. Icefish rely on dissolved oxygen in the plasma. To get enough oxygen to the tissues, icefish have a massive heart that keeps large amounts of blood at a very high pressure.

the blood plasma. About half of the rest binds to hemoglobin, and the remainder—converted into carbonic acid in red blood cells—is in the form of ions: hydrogen ions and bicarbonate ions. Hydrogen ions bind to the protein part of hemoglobin; bicarbonate ions are transported out of the red blood cell into the blood plasma. Both sets of ions are carried to the lungs; there, carbonic anhydrase re-forms them into carbon dioxide, which diffuses away.

A riot of colors

Vertebrate hemoglobin is contained within red blood cells, but in many other groups, such as annelid worms, hemoglobin drifts freely in the blood. In the animal kingdom there are many other respiratory pigments besides hemoglobin. Tunicates, also called sea squirts, are invertebrate chordates and are close relatives of vertebrates. However, they use a respiratory pigment that differs radically from hemoglobin. The pigment contains the metal vanadium. Contained within blood cells called vanadocytes, the pigment gives the blood a bright green color.

The copper-containing pigment hemocyanin occurs in the bloodlike hemolymph of many mollusks and in arthropods such as crustaceans and arachnids. Hemocyanin-containing blood is blue when it is carrying oxygen and colorless otherwise. Hemocyanin is always free in the blood and not contained within cells (unlike vertebrate hemoglobin, which is contained in

red blood cells). However, hemocyanin is only about one-quarter as efficient a carrier of oxygen as hemoglobin.

Many respiratory pigments, like hemoglobin, contain ions of the metal iron. However, the colors of the pigments can be very different. For example, chlorocruorin is a large molecule that is found in the blood of tube-building polychaete worms, such as *Sabella* fan worms. Like hemoglobin, chlorocruorin is deep red when oxygen-laden, but it becomes green when it is deoxygenated.

▲ *Some fan worms have an iron-containing respiratory pigment called chlorocruorin. This pigment, like hemoglobin, is red when carrying oxygen, but it turns green when it gives up the oxygen.*

Ventilation

CONNECTIONS

COMPARE
ventilation in land mammals, such as the **RHINOCEROS**, with that in bony fish such as the **SAILFISH**. Mammals use their ribs and diaphragm to draw air into and out of the lungs, whereas in bony fish water enters the mouth and is pumped or pushed across the gills, which pick up oxygen from the water.

Without a diffusion gradient, gas exchange cannot take place. So animals need to circulate fresh water or air continually over the respiratory surfaces and move out the air that is poor in oxygen and rich in carbon dioxide. This circulation process is called ventilation.

For some organisms, shape alone provides the necessary ventilation without the need for any extra energy expenditure. Mountain midges are flies that pupate (become adult) in streams. The pupae have a shape that creates vortices, or swirls, in the current. The vortices cause a decrease in pressure that makes oxygen collect as bubbles on the insects' gills.

Sponges are riddled with pores; they join to form a central chamber that empties at the osculum, a large hole at the top of the animal.

The shape of a sponge allows it to exploit a physical property of fluids to ventilate itself for free. Fast-moving water creates a lower pressure than slow-moving water. This phenomenon is called the Bernoulli effect. Water moves more slowly at the ocean floor, owing to friction. The water even a few inches above the ocean floor, where the osculum empties, moves slightly quicker. The lower pressure draws water up through the sponge's pores to the osculum. Sponges generally do not live in still waters; they need some water movement to power this respiratory current.

Some land animals also rely on the Bernoulli effect to ventilate their burrows. Prairie dogs, for example, build vast networks of tunnels with openings at different altitudes. Air is sucked through the higher vents from the lower ones to ventilate the network.

Ventilating gills

Most animals must create their own ventilation currents. Crustaceans such as amphipods and many burrowing shrimp flap segments at the end of the abdomen called pleopods to ventilate their gills. Lugworms bring fresh water into their burrows by passing rhythmic waves along their body. Bivalves have banks of tiny filaments called cilia on their ctenidia (gills); they beat back and

▶ **VENTILATION IN SPONGES**
Sponges are covered with pores called ostia, through which water enters a central chamber before exiting through the osculum at the top of the sponge. Pressure is lower in the faster-moving water around the osculum than at the base of the sponge, where the water moves slowly. The pressure difference creates a constant stream of water that ventilates the inside of the sponge.

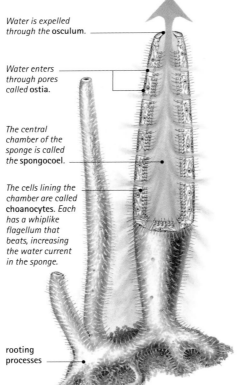

Water is expelled through the osculum.

Water enters through pores called ostia.

The central chamber of the sponge is called the spongocoel.

The cells lining the chamber are called choanocytes. Each has a whiplike flagellum that beats, increasing the water current in the sponge.

rooting processes

COMPARATIVE ANATOMY

Ventilating trachea

Insects need to ventilate their tracheae, but these are kept rigid by taenidia and are largely incompressible. However, there are sections called air sacs that are not stiffened by cuticle. Compression of these air sacs flushes the system. To compress the air sacs, many insects reduce the volume of the abdomen. This can be done by telescoping the segments or by squeezing the upper surface downward. Either way, the pressure of the hemolymph inside rises, squashing the air sacs.

Ventilating gills at speed

Most sharks ventilate their gills in a fashion similar to teleosts, although water may enter through both the mouth and the spiracle, the modified first gill chamber. Water is then forced across the gills before passing out through the gill slits. These sharks can gather oxygen continuously. Some, such as the great white shark, and many fast-swimming teleosts such as mackerel and tuna, do not waste energy by pumping water through the gills. Instead, these fish rely on moving fast relative to the water. Fast movement forces water through the gills, a process that biologists call ram ventilation. Ram ventilators must swim constantly to exchange gases efficiently.

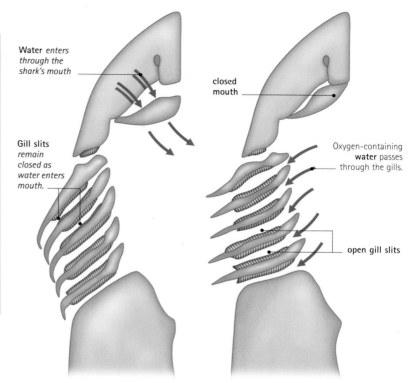

Water *enters through the shark's mouth*

closed mouth

Gill slits *remain closed as water enters mouth.*

Oxygen-containing **water** passes through the gills.

open gill slits

▲ VENTILATING GILLS
Shark (cross section of lower jaw)
Water enters the shark's open mouth (left) when the gills are closed. The shark then closes its mouth and opens the gill slits, allowing the oxygen water to pass across the gills and out through the gill slits (right).

forth to create a current. Cephalopods such as octopuses use oxygen more rapidly than most other aquatic invertebrates and require stronger currents. Cephalopods rely on muscles to open and close the mantle cavity, which is divided into inhalant and exhalant chambers. The muscular activity draws water through a slit, over the octopus ctenidia, and out through a funnel.

Bony fish, or teleosts, such as goldfish also actively pump water over their gills, again, in one direction. Water is drawn into the mouth, which then closes to force it into the gill chambers. They force it across the gills and out. A flap of tissue called the operculum covers the gills, protecting them and closing them off when the mouth is open.

Ventilation in amphibians

Gills are ventilated in one direction, but lungs are ventilated tidally, with air moving in and out along the same path. Lung ventilation is powered in a variety of ways. Amphibians breathe through their nostrils. When the nostrils are open, a flap of tissue called the glottis covers the entrance to the larynx. The floor of the mouth pushes down to suck air into the mouth. The glottis then opens and the lung walls contract, forcing depleted air into the mouth. The nostrils close, and the floor of the mouth rises to force air back into the lungs.

This system may not seem efficient, since air from the lungs mixes with fresh air coming in. However, between cycles the amphibian keeps the glottis closed and nostrils open, and moves the floor of the mouth up and down. This action removes the last of the depleted air from the last cycle. Also, since they have a moist skin, almost all of the carbon dioxide produced by amphibians can be removed through the skin. Exhaled air contains much less of this gas.

Ventilation in reptiles

Reptiles cannot get rid of carbon dioxide through their waterproof skin. Therefore, they must change the air in their lungs frequently to remove as much carbon dioxide as possible. Gas exchange takes place in the front part of a reptile's lungs; the hind part acts as a bellows. Most reptiles breathe in by expanding the rib

Ventilation in mammals and birds

Mammals fill and empty their lungs using a combination of muscles. To breathe in, the diaphragm and intercostal (rib) muscles contract, swinging the ribs outward and expanding the chest cavity. This action lowers the pressure inside the cavity, so air flows in to inflate the lungs. Breathing out occurs as these muscles relax; it is aided by the elasticity of the lung tissue, which contains molecules called elastins that help the tissue recoil. Abdominal muscles also contract to force more air out of the lungs during exercise.

Birds have a very different respiratory system from mammals. When a bird inhales by raising its sternum (breastbone) air fills the lungs and also a system of air sacs. As the bird exhales, oxygen–rich air from the air sacs is forced through the bird's lungs.

▼ VENTILATION IN HUMANS
On inhalation, the diaphragm and rib muscles contract, expanding the rib cage and drawing air into the lungs. The muscles then relax, contracting the rib cage, and air is exhaled.

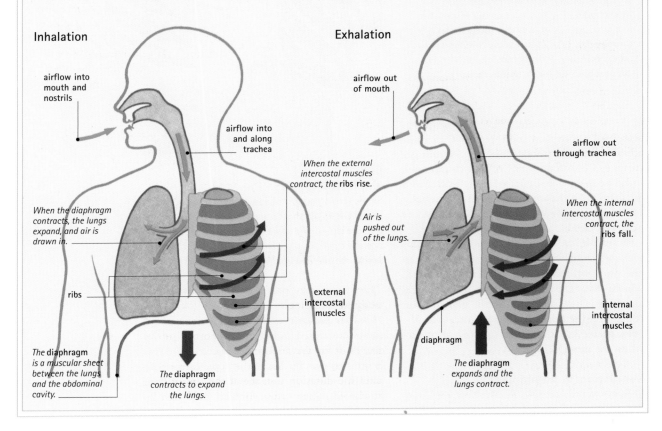

Inhalation

airflow into mouth and nostrils

airflow into and along trachea

When the external intercostal muscles contract, the ribs rise.

When the diaphragm contracts, the lungs expand, and air is drawn in.

ribs

The diaphragm is a muscular sheet between the lungs and the abdominal cavity.

The diaphragm contracts to expand the lungs.

external intercostal muscles

Exhalation

airflow out of mouth

airflow out through trachea

When the internal intercostal muscles contract, the ribs fall.

Air is pushed out of the lungs.

internal intercostal muscles

diaphragm

The diaphragm expands and the lungs contract.

cage; this expands the lungs, causing a drop in pressure so air flows in. The lung tissues and ribs of reptiles are elastic, and this recoil powers the exhalation of the air.

However, ventilation is different in turtles. Their ribs are fixed in place to form part of the shell; turtles rely instead on a sheet of tissue called a diaphragm that connects to the limbs. Extension of the limbs pulls the diaphragm down, causing air to move into the lungs. Exhalation is then caused by limb retraction. Crocodiles use neither ribs nor a diaphragm but a muscle called the diaphragmaticus. This attaches the pelvis to the liver, which is linked to the lungs. When the diaphragmaticus contracts, the lungs expand, causing them to inflate. Breathing out is caused by the contraction of abdominal muscles that pull the liver forward.

Maximizing efficiency

Organsims have a variety of features that maximize their respiratory efficiency. Skin breathers can increase their surface area relative to volume by becoming long in one direction; in other words, by evolving into a wormlike shape. Folds, bumps, or depressions on the skin increase surface area, too. Gills and lungs follow this principle.

CLOSE-UP

Abandoning the nucleus

Mammalian red blood cells have a unique property that allows them to maximize the amount of oxygen in the blood. Mature red blood cells lack a nucleus, the control center containing genetic material that occurs in all other body cells. This allows more oxygen-carrying hemoglobin to be packed in.

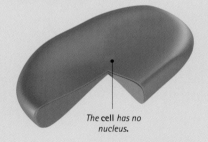

The cell has no nucleus.

▲ **Oxygenated mammalian red blood cell**
Mammalian red blood cells do not have a nucleus. When laden with oxygen from the lungs, the hemoglobin-packed red blood cells are bright red.

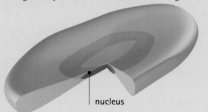

nucleus

▲ **Deoxygenated lizard red blood cell**
Lizards have flatter, more oval, red blood cells with a nucleus. Without oxygen, the iron-containing hemoglobin is blue.

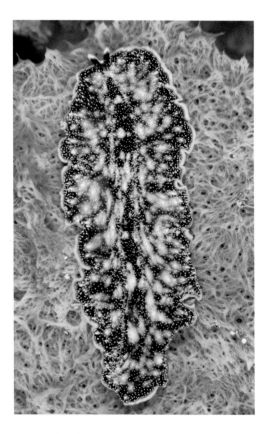

◄ *This marine flatworm does not have any specialized respiratory organs such as gills. It relies solely on the diffusion of oxygen from the seawater across the worm's large surface area. Because the worm is flat, the oxygen does not have to travel far to reach all respiring cells.*

For a skin breather without blood circulation, body parts cannot be situated far from the skin; otherwise, oxygen will not diffuse fast enough to reach them. For example, flatworms and the fronds of the seaweed kelp are long and flat to minimize this diffusion distance. Insects can alter the diffusion distance at the tips of their tracheoles; these tips contain fluid that is withdrawn during periods of activity and returned when the insect is at rest.

Improving the diffusion gradient
The most efficient way to maintain a diffusion gradient is through a countercurrent system, as in fish and cephalopods. Water and blood move in opposite directions; when blood first enters the gill, it is almost devoid of oxygen, but the water it encounters has a little more. As the blood progresses, it encounters richer

water, but there is always less oxygen in the blood than in the water. A diffusion gradient is thus permanently maintained. Teleost gills remove 80 percent of the oxygen in the water that passes through them. Teleosts need such a high efficiency because water contains only about ¹⁄₃₀th of the oxygen of air.

Most lungs are much less efficient than gills. Air moves in and out tidally along the same pathway, so there is a little mixing of fresh and stale air. Only 20 percent or so of the oxygen is removed from inhaled air, but since there is plenty of oxygen available, this small percentage is not a problem. However, birds use oxygen at much faster rates than other land vertebrates. They have a crosscurrent system;

air moves perpendicular (at 90 degrees) to the blood flow. This system in birds is almost as efficient as fish gills.

Saving water

For oxygen to diffuse, respiratory surfaces must be moist. That requirement can cause problems for land animals. Amphibians must remain close to pools or streams for that reason. Most amphibians that live in arid areas, such as spadefoot toads, spend much of their lives in a state of inactivity called estivation. An exception to the rule that moist-skinned amphibians living in arid areas must remain inactive for most of the time is the Gran Chaco monkey tree frog: it lives in dry areas,

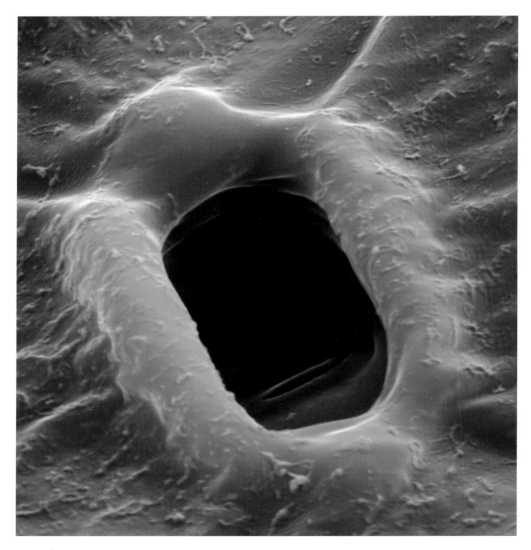

▶ The stomata, or air pores, of certain desert plants, such as aloe vera, open only at night. Thus they minimize the loss of water through the stomata during the hot day. In these plants, called CAM plants, the carbon dioxide collected at night is stored as an acid, before being broken and used to form sugars during the day (magnified 2,000 times).

but secretes a waxy substance that it wipes over its skin to resist water loss.

Land animals such as reptiles and insects have waterproof skin and have the ability to shut off their respiratory surfaces. Mammals and birds breathe relatively rapidly, so potentially more water can be lost. They solve this problem with a series of bony or cartilaginous ridges called nasal turbinates. These ridges run from the back of the nose to the epiglottis and are richly supplied with blood vessels. The nasal turbinates warm

▲ *Atlantic spotted dolphins "porpoise" when swimming fast. Leaping out of the water allows them to breathe efficiently without slowing down.*

incoming air and moisten it to prevent the airways from drying. They then recapture water vapor as it is breathed out.

Efficient ventilation at speed

Many mammals do not actively breathe when they are running—the power of movement does the job for them. Once a horse breaks into a gallop, its breathing and stride patterns coincide closely. As it extends its forelegs, the rib cage and viscera, or internal organs, shift forward, pulling the diaphragm and causing inhalation. When the forelegs touch down, the viscera shift back and the rib cage is compressed, forcing air from the lungs.

Swimming birds and mammals also optimize breathing efficiency. Penguins and dolphins often "porpoise" at high speed: they leap through the air as they swim. Swimming part-submerged at the surface uses more energy than swimming underwater. Porpoising allows the animals to breathe without slowing down and minimizes the energy used for swimming.

JAMES MARTIN

IN FOCUS

How plants save water

Plants that live in dry areas fix carbon dioxide into sugars in a different way from other plants to help them save water. Most plants use a process called C3 photosynthesis, so named because carbon dioxide is first fixed into a molecule with three carbon atoms. These plants keep their stomata (air pores) open all day long. Plants in dry areas use a different process, called C4 photosynthesis; they first incorporate carbon dioxide into a four-carbon molecule. This process is much quicker than C3 photosynthesis, so long as the weather is hot and bright, and the stomata can be closed for longer periods in the day. Plants such as corn are C4 photosynthesizers. Desert species such as cacti are called CAM, or crassulacean acid metabolism, plants. Cacti stomata open only at night when it is cool, so little water is lost. Carbon is stored in an acid, which is broken down for incorporation into sugars during the day.

FURTHER READING AND RESEARCH

Kardong, Kenneth V. 2005. *Vertebrates: Comparative Anatomy, Function, Evolution.* McGraw-Hill Education: New York.

Raven, Peter H., George B. Johnson, Susan R. Singer, and Jonathan B. Losos. 2004. *Biology.* McGraw-Hill Science: New York.

Skeletal system

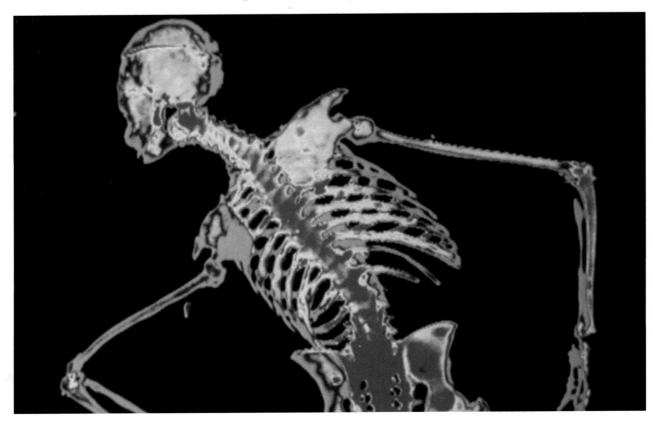

▲ *The tough, flexible human skeleton has about 206 bones of various shapes and sizes. This image shows the bones of the skull, vertebral column, ribs, pelvic girdle, and arms.*

Think of skeletons, and it is difficult to escape the concept of something dead. The supportive structures—bones and other hard parts, like shells—of many animals are the most enduring components of the body and often survive long after the animal is dead, when the flesh and other soft tissues have rotted away. However, in a living organism skeletal structures can be far from inert. They are very much living tissues, which change in response to use and within which a number of vital biochemical processes take place.

The skeleton is the part of an organism's body that gives it strength, rigidity, and support. In conjunction with muscle it makes precise movement possible, and in most animals the skeleton offers protection for other soft, vulnerable parts of the body such as the brain, heart, lungs, and other vital organs.

EVOLUTION OF THE SKELETON Hard skeletons evolved about 500 million years ago in sea animals. *See pages 171–174.*

PLANT SUPPORT STRUCTURES The vascular system of plants (the xylem and phloem) also provides support. *See page 175.*

INVERTEBRATE SUPPORT STRUCTURES Many invertebrates have a hydrostatic skeleton, some have an exoskeleton, and others have an endoskeleton. *See pages 176–181.*

VERTEBRATE SKELETON Most vertebrates have a skeleton made of bone. *See pages 182–183.*

VERTEBRATE SKULL The skull is the skeleton of a vertebrate's head. *See pages 184–185.*

VERTEBRATE AXIAL SKELETON The skull, vertebral column, and rib cage. *See pages 186–187.*

VERTEBRATE APPENDICULAR SKELETON The legs, wings (in birds and bats), and pelvic and pectoral girdles make up the appendicular skeleton. *See pages 188–189.*

Evolution of the skeleton

Life began in the sea. Early organisms were tiny: they were single cells surrounded by a membrane that held together all the vital components of life in gelatinous soup called the cytosol. In most such organisms, the membrane was reinforced with a network of microscopic fibers and an underlying scaffold of tubules. These networks gave cells a means of support and helped generate movement—the creeping of amoebas and the swimming of other protists. In this sense, the fibers and tubules were forerunners of both muscular and skeletal systems.

However, true skeletons did not evolve for a very long time. For much of the history of life on earth, organisms had no rigid support structures of any kind. Life-forms were fully aquatic and relied on water to support their bodies. Water is about 800 times more dense than air, so the organisms could function without a rigid skeleton. The fossil record for these early life-forms is very sparse precisely because they had very few hard parts that could be preserved.

This soft-bodied state of affairs lasted millions of years. In the absence of skeletons or any other hard parts, animal life nevertheless achieved an amazing level of diversity by 550 million years ago. However, even with water for support, a soft body has limitations, most notably a lack of power and speed. These are attributes that create a huge advantage for predators and prey alike, but they are

Fossilization

Fossil evidence is enormously valuable in tracing the progress of evolution, but the discovery of useful fossils depends on a long sequence of improbable events, and so the record is very patchy. Fossils form when a organism dies in a place in which the body can lie undisturbed and be rapidly covered with fine sediment. Lake beds are ideal. As more and more sediment is deposited on top, the layers in which the remains lie begin to turn into rock. Gradually, the hard body parts (that is, the skeleton) are mineralized—the original organic compounds are replaced with inorganic ones. After millions of years, the bone is turned to rock, but the fossil will be found only if geological processes bring the rock in which it lies back to the surface where erosion may eventually expose it.

dependent on muscle power. And for muscles to operate efficiently and forcefully, they need something rigid to pull against.

About 500 million years ago, there was a period of great expansion in the number and variety of life-forms on Earth. During this period, called the Cambrian explosion, some animals developed hard skeletons that allowed them to maintain their body shape and move quickly. That offered a big advantage to predatory animals—and to prey animals trying to flee from predators. Thus skeletons and other support structures played an important part in a kind of evolutionary arms race. Once skeletons appeared, they evolved rapidly. Thus

▼ CROSS SECTION OF THE DIATOM NAVICULA
All the living matter of a diatom—a type of unicellular organism—is contained within the frustule, a rigid exoskeleton that is impregnated with silica.

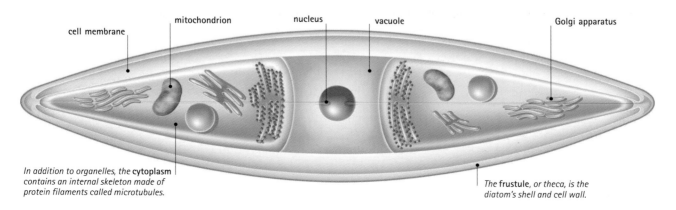

cell membrane   mitochondrion   nucleus   vacuole   Golgi apparatus

*In addition to organelles, the **cytoplasm** contains an internal skeleton made of protein filaments called microtubules.*

*The **frustule**, or theca, is the diatom's shell and cell wall.*

CONNECTIONS

COMPARE the skeleton of a **COELACANTH** with that of a **CROCODILE**. Similarities suggest how the skeleton of lobe-finned fish, such as the coelacanth, was modified for walking on land.

COMPARE the skeleton of an **OSTRICH** with that of an **EAGLE**. The ostrich evolved from a flying bird, so its forelimbs are more like those of birds than like those of other vertebrates that walk on two legs.

the transition from virtually no fossils in rocks more than 550 million years old to rocks full of an incredible diversity of fossils occurred relatively quickly.

The animals of the Cambrian explosion include representatives of most of the major animal groups that are extant (alive) today, plus many that have since become extinct. Alongside the ancestors of echinoderms, mollusks, and arthropods were small creatures with a long, bilaterally symmetrical body that were to give rise first to fish and then to other complex vertebrates.

The first hint of a skeleton in these simple pre-vertebrates was a long, sausage-shape structure called the notochord running along the animal's back. It was encased in a tough sheath that gave it rigidity and helped hold the body in shape, preventing the softer tissues from overstretching or being telescoped. The fish with the most primitive features alive today (hagfish and lampreys) still depend on a notochord for support, and the same structure is present in the embryos of all vertebrates. As

fish evolved, the notochord came to be replaced by a backbone made of articulated cartilage or bone structures. Having evolved the ability to produce these sturdy skeletal materials, fish evolved to provide support and protection elsewhere in the body: in the head and fins.

From fins to limbs

Vertebrates with limbs instead of fins are called tetrapods. Tetrapod means "four feet," though the description applies to many animals in which the feet are modified into something else, such as flippers or wings; and to some that have no legs or feet at all, such as snakes.

Scientists agree that the limbs of tetrapods evolved from the fins of certain types of fish called sarcopterygians. There is less agreement about exactly how this development took

▼ Priscacara liops *is an extinct lobe-finned fish that lived in North America 50 million years ago. All land vertebrates evolved from an ancestral lobe-finned fish, with the bones of the fish's fleshy pelvic and pectoral fins developing into the limb bones.*

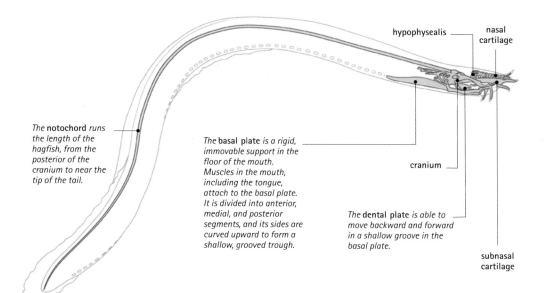

hypophysealis — nasal cartilage

The **notochord** runs the length of the hagfish, from the posterior of the cranium to near the tip of the tail.

The **basal plate** is a rigid, immovable support in the floor of the mouth. Muscles in the mouth, including the tongue, attach to the basal plate. It is divided into anterior, medial, and posterior segments, and its sides are curved upward to form a shallow, grooved trough.

cranium

The **dental plate** is able to move backward and forward in a shallow groove in the basal plate.

subnasal cartilage

◀ **Atlantic hagfish**
The skeleton of a hagfish is made of cartilage, not bone. Most of the elements of the skeleton are in the head. Unlike that of other adult vertebrates, the spinal cord of a hagfish is protected not by bony vertebrae but by a flexible notochord. A hagfish lacks ribs.

place, but the sequence of events was probably something like this. An ancestral fish had two pairs of lobed fins, like those of modern-day lungfishes or coelacanths. The ancestor was able to use these fins as levers to help pull itself along the sea floor. In shallow water, where the fish's body weight was not fully supported by the water, the technique was more effective than conventional swimming, and by pressing its fins into the ground as it wriggled from side to side the fish was even able to "swim" on land. At first, the fin-limbs did not carry much weight, but as fish found new advantages to excursions onto land (finding food and avoiding predators), natural selection began to favor those with longer and stronger limbs and pectoral and pelvic girdles (shoulders and hips) sturdy enough to take the strain. Thus evolved the first amphibians.

Taking the strain
The backbone of a fish acts to maintain the shape of the body, in particular resisting compressive forces and keeping the animal from telescoping or overstretching. In most species of fish, the backbone also has long spines for the attachment of the large trunk muscles that power swimming. However, since terrestrial vertebrates move on four legs without the support of water, their backbone also has to take on a weight-bearing role.

Different branches of the vertebrate family tree have adopted different approaches to the problem of moving and supporting the body on land. Salamanders have stuck with the original solution. Their skeleton has many similarities to that of their fish ancestors, and they retain a fishlike side-to-side wriggle when swimming and when moving on land. A salamander's legs are short and weak. When the salamander is resting, the weight of the body is lowered onto the ground. Lifting the body creates enormous stress on the legs and the long backbone, and thus walking is tiring and inefficient for these animals. In frogs and toads, walking is much less difficult, mainly because the body is much shorter. A typical frog has just nine relatively large vertebrae, and they interlock closely to form a rigid support with virtually no side-to-side flexibility. The sacral vertebrae are fused to the pelvic girdle, which is strengthened by a rigid central rod of bone,

EVOLUTION

The first tetrapod?

The earliest known tetrapod is an amphibian called *Ichthyostega*, which lived about 360 million years ago. Apart from the limbs, the skeleton of *Ichythostega* is very much like that of a fish. *Ichthyostega* would have spent much of its time in water, but it was also able to move (and breathe) on land.

Archaeopteryx: A missing link

There is no doubt that birds evolved from reptiles. However, when fossil hunters discovered the proof, it surpassed all expectations. *Archaeopteryx* is a classic example of a missing link—an extinct organism showing a blend of characteristics from two groups and demonstrating how one group might have evolved from the other. *Archaeopteryx* had the skeleton of a reptile, with a long tail, small sternum (breastbone), small coracoid, and five-fingered forelimbs. But clearly, it also had feathers.

▼ *Many scientists believe that birds evolved from dinosaurs about 150 million years ago. This is the most complete fossil of Archaeopteryx, which was found in Solnhofen limestone deposits in Germany in 1880 and is now housed in the Humboldt Museum, Berlin, Germany.*

Toe count

Fossil evidence shows that early tetrapods had a variable number of digits on each limb: for example, eight in *Acanthostega*, seven in *Ichthyostega*, and six in *Tulerpeton*. However, in the evolutionary line that gave rise to more advanced vertebrates, five digits were the norm. The hands, feet, flippers, wings, and paws of all living amphibians, reptiles, birds, and mammals are derived from the same five-fingered structure, known as the pentadactyl limb.

Reptiles

Reptiles show the greatest skeletal diversity of any group of vertebrates. They have varied in size from tiny 1-inch-long geckos to colossal dinosaurs such as *Diplodocus* and *Brachiosaurus*, which stood up to 60 feet (18 m) tall and weighed 60 tons (61 metric tonnes). To put this into perspective, you could crush the whole skeleton of the smallest lizard with one toe, but just one neck bone from the largest known dinosaur could do the same to you!

Reptile skeletons vary in form as well as size. Faced with the skeletons of a large turtle and a python, it is difficult to see how one, with its short body, beaklike jaw, flippers, and shieldlike carapace, can possibly be related to the other, with its loose, spindly jaw, enormously long backbone, hundreds of ribs, and absence of functional limbs.

Mammals and birds

Compared with the very great diversity of the reptiles, the skeletons of mammals and birds, both of which groups are descended from reptilian ancestors, are relatively uniform. The body plan of birds, in particular, is restricted by the requirements of flight. Even flightless birds like the ostrich and kiwi are constrained by this, since they evolved from ancestors that flew. Birds have fewer skeletal elements than either reptiles or mammals because many bones are fused to provide the strength and rigidity needed for flight.

the urostyle, derived from fused caudal vertebrae. The pelvic girdle of frogs is very large because it connects the backbone to the extremely long hind legs.

Plant support structures

We tend not to think of plants as having skeletons, but they do have support structures. Often these structures are so effective that they last for hundreds of years. The skeletal material produced by the largest vascular plants—the wood of trees—is equally useful to humankind for construction and as fuel.

Nonanimal life, just like that of animals, began in the seas with the appearance of single-cell organisms about 1 billion years ago. Over time, some of these very simple life-forms gave rise to colonial and multicellular species, which had the potential for division of labor and specialization of different cell types. As with animals, plants had little need of rigid support as long as they remained in the water, and even after they encroached onto land many stuck with a simple form that simply encrusted or draped over existing structures and required little support. For aquatic plants, however, there are big advantages to growing upward, reaching toward the sun and pushing reproductive structures into the air where spores and seeds are more likely to be dispersed by the wind.

Plant skeletons evolved in tandem with the internal transport systems of vascular plants. In fact, to a large extent skeleton and vascular systems are one and the same thing. The branching skeleton seen in the leaves of vascular plants is a network of tiny xylem fibers and phloem tracheids that carry water and nutrients to every part of the plant. The vessels extend along every stem and twig, through the trunk of trees and into the roots. Because the walls of these tubes are reinforced to keep them open, they offer support to the plant as a whole. The supporting role of tracheids is enhanced by a thick layer of cells and associated material called the cambium. The cambium of woody plants is particularly large, and contains fibers of complex carbohydrates called cellulose and hemicellulose, held together with a special protein called lignin.

CLOSE-UP

Tendrils

Many climbing plants, such as sweet peas, cucumbers, and passionflowers, have threadlike structures called tendrils for support and attachment. Depending on the species of plant, tendrils can be modified leaves, stems, or petioles (leaf stalks). On contact with a solid object, such as a branch, a tendril twirls around it—a phenomenon known as thigmotropism. The cells that make contact with the object lose water and become smaller than the outer cells. In this way, the tendril curves around the object, supporting the plant as it grows upward.

▼ TRUNK CROSS SECTION

Apple tree

The outside layer of the trunk is nonliving bark, which provides a support structure for the tree. Beneath the "skeleton" of the trunk is the cambium, which transports water and lays down new xylem and phloem tissue.

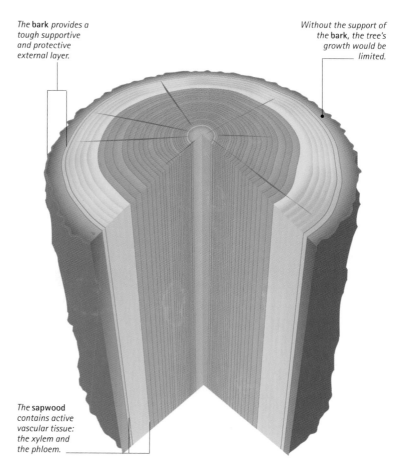

*The **bark** provides a tough supportive and protective external layer.*

*Without the support of the **bark**, the tree's growth would be limited.*

*The **sapwood** contains active vascular tissue: the xylem and the phloem.*

Invertebrate support structures

CONNECTIONS

COMPARE the exoskeleton of the *WEEVIL* with that of the *CRAB*. The weevil cuticle is less mineralized than that of the crab because it does not have access to such a ready supply of calcium carbonate.

COMPARE the hydraulic support system of the *EARTHWORM* with that of the *SEA ANEMONE*. These animals are not related, but both use water pressure to give their body rigidity and to move about.

Even the very simplest animals can have skeletons. Sponges are the simplest form of multicellular animal life. The body has just two layers (compared with three in most multicellular animals). Between the inner and outer surfaces of the sponge body is a mass of supporting material called mesohyl. In most sponges, this is a gel-like mass containing various stiffening elements such as protein fibers (collagen and spongin) and mineral spicules. The stiffening elements are sometimes preserved when the animal dies, leaving a delicate, brittle skeleton that looks like fine lace.

Even animals that have no hard body parts must be able to hold their body shape. Many soft-bodied organisms use water pressure to achieve this. To understand how water can help give a structure form and rigidity, imagine an empty balloon. It contains no hard parts and on its own is flimsy and floppy. However, if the balloon is filled with water and sealed, it takes on a three-dimensional form. The more water it contains, the tauter the latex of the balloon is stretched and the more rigid the shape becomes. All living organisms gain some measure of support from the water they contain. Plants are no exception: they wilt and loose rigidity when dehydrated.

Water cannot be compressed. It resists being squeezed by flowing from the areas of greatest pressure to areas of less pressure. Just as a hard skeleton provides muscles with something to pull on, tissues that are firmly plumped up with water can offer the resistance required to generate muscular locomotion. Zoologists call this rigidity a hydraulic skeleton, and soft-bodied animals use it to move. By contracting

▶ The seawater-filled cavity (coelenteron) of jellyfish, such as the many-ribbed hydromedusa (right), and its gel-like body tissues form a supportive structure called a hydrostatic skeleton.

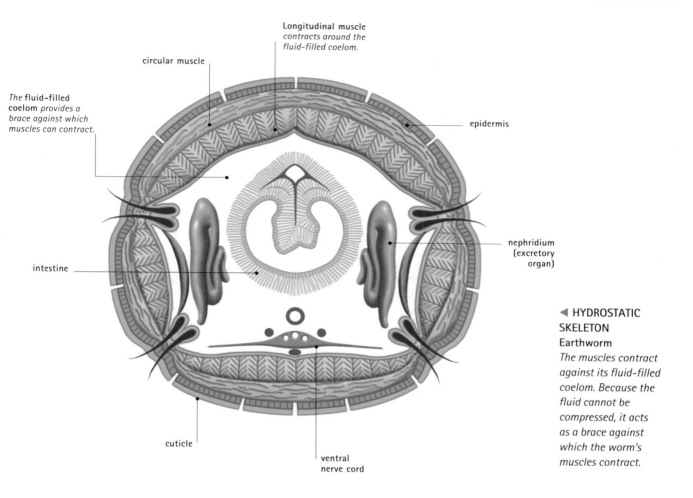

circular muscle

Longitudinal muscle
*contracts around the
fluid-filled coelom.*

*The fluid-filled
coelom provides a
brace against which
muscles can contract.*

epidermis

intestine

nephridium
(excretory
organ)

cuticle

ventral
nerve cord

◄ HYDROSTATIC
SKELETON
Earthworm
*The muscles contract
against its fluid-filled
coelom. Because the
fluid cannot be
compressed, it acts
as a brace against
which the worm's
muscles contract.*

muscles in one part of the body, an animal such as a sea anemone is able to force water into another area, such as a tentacle, which must then expand to accommodate the water.

The hydraulic skeleton is developed to best effect in the annelids, segmented worms such as earthworms, ragworms, and leeches. Each segment in the annelid body is effectively a bag of fluid wrapped in several thin layers of muscle. This arrangement allows the animal to move in a variety of ways. For example, when the muscle on one side of a segment contracts, the other side expands to accommodate the fluid. When this happens simultaneously in several segments, the whole body curves toward the side that is contracting. By alternating contractions the worm performs a characteristic wiggle that can propel it forward; that is how ragworms crawl and swim.

IN FOCUS

A change of armor

The rigid exoskeleton of crustaceans such as lobsters and crabs is not elastic, and in order to grow the animal must shed its protective covering and produce a new, larger one. The process may have to be repeated dozens of times during the animal's lifetime, and the process becomes increasingly arduous as the animal grows. Before shedding its armor, the animal becomes dehydrated, so that it shrinks a little inside its armor. It also reabsorbs minerals from the exoskeleton into the body. The exoskeleton becomes brittle and splits open, allowing the animal to crawl out. To begin with, the new exoskeleton is soft and crumpled, but as the animal swells back to normal size, it straightens out and begins to harden. The process can take several days, during which time the soft-bodied animal is very vulnerable to predation. The animal often eats its old exoskeleton to help replace lost protein and minerals.

Alternatively, if circular muscles cause a segment to constrict, hydraulic pressure forces it to lengthen. Thus a worm can make each segment (and its whole body) short and fat, or long and thin. By alternating from one state to the other, annelids like earthworms effect a gradual creeping type of locomotion.

The echinoderm test

Echinoderms are a group of marine animals including starfish, sea cucumbers, and sea urchins. Most sea cucumbers are soft-bodied, but all other echinoderms possess some kind of internal skeleton.

In starfish the skeleton is made up of a lattice of mineral rods, crosses, or plates embedded in a layer of connective tissue within the body wall. In species such as the common red sea star, the skeleton is relatively loose, whereas in other species it can be very highly mineralized and brittle. In sea urchins, the skeleton is even more developed; it is a regular jigsaw of plates

that form a roughly spherical or discus-shaped shell called a test. The plates of the test are covered with small bumps, to which the mobile spines are attached in the living animal. The plates are perforated by tiny holes through which the suckerlike tube feet can be extended when the animal crawls. The tube feet operate by water pressure, a little like the hydraulic skeletons of corals, sea anemones, and annelids.

The shells of mollusks

Mollusks are essentially soft-bodied animals that have developed a characteristic form of body armor. The shells, or valves, of mollusks such as mussels, clams, snails, and limpets are not really skeletons, but they protect the animal from physical attack and—in many tidal and land-dwelling species—from dehydration. The shells may also serve as a scaffold from which some species can suspend organs such as gills and filters, which lack their own rigidity and would not work if allowed to

▼ *Many mollusks, including the queen conch (below), have a hard mineralized shell. Although not a skeleton, the shell protects the animal and may be a rigid attachment site for internal organs.*

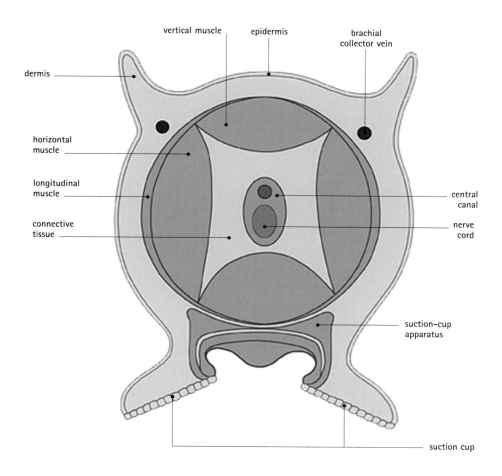

vertical muscle
epidermis
brachial
collector vein
dermis
horizontal
muscle
longitudinal
muscle
connective
tissue
central
canal
nerve
cord
suction-cup
apparatus
suction cup

◄ CROSS SECTION OF
OCTOPUS TENTACLE
*An octopus does not
have an internal
skeleton. Its arm and
sucker are formed from
a complex arrangement
of muscles, working
against one another
and enabling the
octopus to wrap around
and grip very tightly
onto prey.*

lie flat. In cuttlefish and squid the shell is internal; in octopuses it has disappeared.

The tissues of a mollusk's mantle—the soft part of the animal that contacts the shell directly—lay down the shells throughout the animal's lifetime. The shells of mollusks are made of alternating layers of protein and two different crystalline forms of calcium carbonate: calcite and aragonite. Marine mollusks have access to much more calcium carbonate than freshwater or land-dwelling species and thus can create heavier shells. The shiny material lining the shells is called nacre, or mother-of-pearl. It contains about 95 percent aragonite, but the bulk of the shell matrix in marine mollusks is usually calcite, which is more stable chemically in seawater.

CELL BIOLOGY

Arthropod cuticle

The exoskeleton of arthropods such as insects and spiders is made of protein and chitin, a polysaccharide secreted by the cells in the underlying skin. The cuticle is made of several layers: a thin, waxy, outer epicuticle; a pigmented exocuticle; and a thicker endocuticle. The exocuticle is missing from the jointed areas, making the exoskeleton flexible. In large crustaceans such as crabs and lobsters, the exocuticle and endocuticle also contain large quantities of minerals, especially calcium carbonate and calcium phosphate.

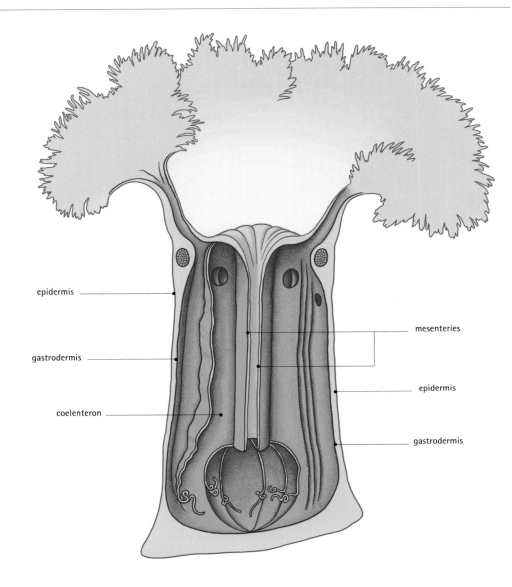

**► Common plumose
sea anemone**
*Two walls of cells—
the epidermis and
gastrodermis—provide
a support structure
around the animal's
body cavity, or
coelenteron.
Internal walls called
mesenteries give
additional support.*

epidermis

gastrodermis

coelenteron

mesenteries

epidermis

gastrodermis

Arthropod exoskeleton

The arthropods are the largest and most
successful group of invertebrates and include
insects, arachnids, centipedes, millipedes, and
crustaceans. The group owes much of its
remarkable success to its unique suit of armor,
the exoskeleton. The exoskeleton consists of a
cuticle made of the polysaccharide chitin and
serves as both structural support and body
protection. In marine arthropods, the cuticle
is often reinforced with deposits of calcium
carbonate and calcium phosphate. To allow
movement, the cuticle is jointed at flexion

points all round the body where the cuticle is
thin and bendy. In "primitive" arthropods, such
as wood lice, millipedes, and the extinct
trilobites, the body plates follow the same
pattern as the body segments, with each
segment encased in four plates: the tergum on
the back, the sternum on the underside, and
two side plates called pleura.

In other groups of arthropods, such as insects
and arachnids, the segmental arrangement is
sometimes lost over part or all of the body, and
the plates are either fused or subdivided. The
legs and other appendages are sheathed in

chitin tubes joined by articulating membranes that allow them to bend. The word *arthropoda* means "jointed feet."

Growing pains

One of the problems with living inside a rigid structure is that it restricts an organism's ability to grow. Arthropods get around this problem by shedding their exoskeleton periodically in a process called ecdysis, or molting. Mollusks and some single-cell organisms called foraminiferans molt by laying down more material at the edges of their shell, resulting in the ever-enlarging spiral of a snail shell or the concentric growth rings on the valves of mussels and scallops. For an echinoderm such as a sea urchin, in which the test must grow with the animal, shedding the test and adding material at the opening are not practical solutions. Instead, each of the plates in the test grows around its edges. Plates that are actively growing are only weakly attached to

Coral reefs

Coral polyps would be inconspicuous animals were it not for the enormous support structures they build. These amazing structures, which accumulate to form vast reefs, are more like houses than skeletons, but the stony material from which they are made is secreted by the tiny soft-bodied animals living within.

their neighbors. Thus the test of a sea urchin that dies during a period of growth falls apart easily. A sea urchin that has not grown much recently, perhaps because of a lack of food, has a stronger test in which the joints, or sutures, between the plates are heavily mineralized and firmly attached to one another.

▼ *A lobster's soft parts are enclosed in a tough cuticle, or exoskeleton, which provides support for the animal, points of attachment for muscles, and protection against predators. The cuticle of lobsters is made of chitin, a complex carbohydrate similar in structure to cellulose, bound up with various proteins. The cuticle is much thinner at the joints, to allow movement.*

Vertebrate skeleton

Bone is the predominant skeletal material in most adult vertebrates. Like cartilage, it develops from connective tissue and consists of isolated cells that secrete a resilient matrix material. Whereas the matrix of cartilage is usually springy and translucent, that of bone accumulates mineral salts, mainly calcium carbonate and calcium phosphate, which combine to form a material called hydroxyapatite. The mineralization of bone

Osteoblasts and osteoclasts

The bones of vertebrates are laid down and sculptured throughout life by the actions of two types of bone cells. Osteoblasts secrete bone matrix and remain active even when an animal is fully grown, helping broken bones heal and replacing old bone material that is continually being destroyed by the other important type of bone cell, the osteoclasts. Osteoclasts digest bone matrix, allowing the constituent minerals to be recycled. In adult humans, about 10 percent of bone is replaced in this way each year, ensuring that the skeleton stays strong. In older women, there is a tendency for osteoclasts to demineralize bone faster than osteoblasts can replace it. This demineralization leads to a weakening of the bones—a disorder called osteoporosis.

▼ *This micrograph shows human osteoblasts. These cells are found in bone and secrete substances such as collagen that form bone matrix.*

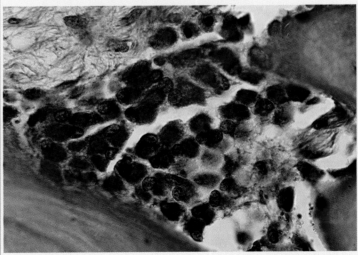

Bone density

Compared with the bones of mammals, those of birds are about one-third the weight. Bird bones are hollow, with an internal scaffold of fine struts, making them strong but light. This characteristic is a big advantage for flight. The bones of some animals are unusually dense and heavy even when compared with mammals. Ocean-dwelling manatees, for example, have heavy bones that help counteract the buoyancy of their thick layer of fat. Heavy bones allow them to feed on underwater plants without having to expend energy to stay submerged. The density of bone is also related to its strength: in humans, the lower jaw is the hardest, densest bone, whereas in male deer the antlers are hardest and densest.

causes it to harden and become opaque. Since bone is impermeable to nutrients, it requires its own blood supply. Thus it is penetrated by a fine network of microscopic channels called Haversian canals that allow blood and lymph to pass to and from the cells embedded in the tissue. The center of a large bone is particularly richly supplied with vessels. There, the bone structure is relatively open and spongy-looking, with many small holes and channels. This spongy area also contains bone marrow, a fatty substance important in the production of blood cells.

The support structures of vertebrates are true skeletons; they are internal scaffolds made of two different kinds of tissues—bone and cartilage. Both types of tissues are derived from connective tissue, but they are very different materials.

Cartilage is a tough, flexible tissue that begins to form early in the development of an embryo. Cartilage performs a supporting role long before bones develop. Usually cartilage is a deep tissue: that is, it does not form close

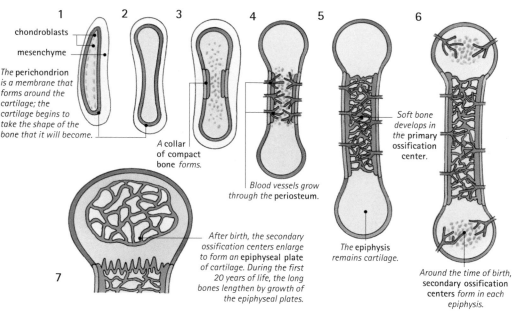

1
chondroblasts
mesenchyme

The perichondrion is a membrane that forms around the cartilage; the cartilage begins to take the shape of the bone that it will become.

2

3

A collar of compact bone forms.

4

Blood vessels grow through the periosteum.

5

The epiphysis remains cartilage.

6

Soft bone develops in the primary ossification center.

Around the time of birth, secondary ossification centers form in each epiphysis.

7

After birth, the secondary ossification centers enlarge to form an epiphyseal plate of cartilage. During the first 20 years of life, the long bones lengthen by growth of the epiphyseal plates.

◀ **DEVELOPMENT OF LONG BONES IN A FETUS AND CHILD Human**
Fetal bones develop from tissue called mesenchyme. The chondroblasts are cartilage-making cells. Diagrams 1–5 show different stages as the embryo grows. 6 illustrates a long bone as it is around the time of birth, and 7 shows the bone of a young child.

◀ *In addition to an internal skeleton, armadillos, such as this nine-banded armadillo, are covered by a "shell" or exoskeleton of bony horn-covered plates that protects the animal from predators.*

under the skin or at the body surface. Exceptions to this general rule include the human nose and the external structure of the ears in most mammals.

Hyaline cartilage, a glassy type of cartilage found in joints, is a slightly elastic tissue in which the predominant compound is a polysaccharide called chrondromucoprotein. This translucent, smooth-textured material is a sort of set gel secreted by specialized cells called chondrocytes. The chondromucoprotein matrix is strengthened by many fibers crisscrossing through it. The chrondrocytes become trapped by the material that they secrete, ending up encapsulated in lacunae, small bubbles within the cartilage. These cells receive all the vital nutrients and oxygen they need to function by simple diffusion through the cartilage gel, since only large cartilaginous structures have a blood supply. Not all cartilage has the same springy texture. In vertebrates such as sharks, whose entire skeleton is composed primarily of cartilage, the tissue is so heavily calcified that it resembles bone.

Vertebrate skull

CONNECTIONS

COMPARE the skull of a **CHIMPANZEE** with that of a **CROCODILE**. The chimpanzee's skull has orbits in the front, so the eyes face forward and provide good binocular vision. The crocodile has a long snout, with the orbits high on the sides of the skull, giving good all-around vision. The chimpanzee's cranium, which houses a large brain, takes up a large proportion of the skull, whereas that of the crocodile is relatively small. The lower jaw of the crocodile is almost as large as the upper jaw, whereas the chimp's lower jaw is far smaller.

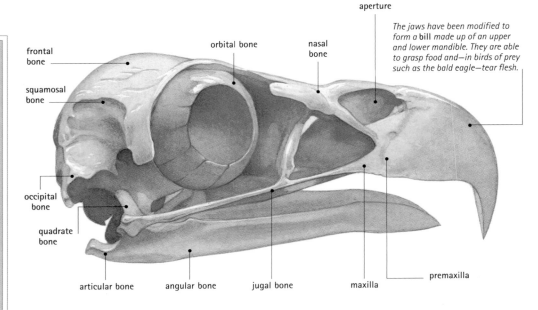

The jaws have been modified to form a **bill** made up of an upper and lower mandible. They are able to grasp food and—in birds of prey such as the bald eagle—tear flesh.

frontal bone · squamosal bone · occipital bone · quadrate bone · articular bone · angular bone · jugal bone · orbital bone · nasal bone · nasal aperture · maxilla · premaxilla

▲ Bald eagle
The skull of a bird is composed of thin sheets of bone with tiny supporting struts. The skull is extremely strong to support the jaw muscles and protect the brain. The jaws of a bird are modified to form a grasping bill. Eagles have a hooked upper mandible that is used to tear flesh.

In all vertebrates the brain is encased in a bony or cartilaginous box called the cranium, which is the most important element of the skull. Despite its appearance, the skull is not a single large bone but a structure formed from many smaller bones, which become fused during the development of the animal. This is more obvious in some animals than in others. In typical mammals, for example, the skull is made up of 34 bones. In a mature skull, the joints between the bones appear as fine, squiggly lines. In most vertebrates the skull can be divided into two main parts: the braincase, or cranium, which includes the facial bones;

and the part that in fish forms the gill arches and in other vertebrates forms the lower jaw.

The skull of both birds and reptiles has a characteristic articulation (joint) of the lower jaw made up of articular and quadrate bones. In mammals, the lower jaw consists of just one bone, the dentary. In the reptile ancestors of mammals, there were three pairs of accessory jawbones: the angular, articular, and prearticular bones. In mammals, these three pairs of bones have become the malleus, incus, and stapes of the inner ear. They are greatly reduced in size but perform an important role in transmitting the vibrations mammals interpret as sound. The acute hearing of birds is due in part to a different arrangement: vibrations are transmitted through a single auditory ossicle (ear bone) called the columella.

In all vertebrate embryos, the development of the skull follows a more or less similar pattern,

CLOSE-UP

Beaks

When the ancestors of birds evolved forelimbs devoted to powered flight, they had to sacrifice many of the other useful functions of forelimbs, such as manipulating food, grooming, signaling, fighting, crawling, climbing, and grasping. In compensation, birds use their jaws for a much greater range of functions than other tetrapods. The jaws of birds are encased in a sheath of horny material, the beak. Bird beaks have evolved an astounding array of shapes and sizes, allowing birds to perform a variety of highly specialized tasks.

Horns and antlers

The impressive headgear of hoofed animals such as cattle, rhinoceroses, and deer is formed largely from bone. Horns are permanent structures that grow from the frontal bones of all male and some female cattle, antelope, and related animals. Horns are sheathed in a layer of tough keratin, the same material that is found in hair and nails. Antlers are temporary structures grown and cast off annually by male deer; antlers have no horny covering.

which repeats the sequence of evolutionary development. The braincase develops from neural crest tissue at the back of the brain. First to appear are a pair of small cartilaginous structures, one on each side of the front end of the notochord. These grow and fuse around the notochord, forming a basal plate, which grows out around the brain, meeting and fusing with

several other newly formed cartilaginous structures until the brain is enclosed in a loose, symmetrical jigsaw of plates. At about the same time, a series of paired cartilaginous struts form in the region of the throat, beneath the skull. In fish, the struts become the gill support arches. The first of these arches, called the mandibular arch, progresses to form part of the upper and lower jaws, but these primary jaws persist only in sharks and rays. In most bony fish and higher vertebrates they are soon replaced by new bone derived from the overlying skin. The second set of cartilaginous structures form the hyoid arch, which in fish helps suspend the jaw from the main part of the skull. In more advanced vertebrates, the hyoid arch supports the base of the tongue and is used in vocalization.

In all vertebrates except cartilaginous fish, the cartilaginous structures of the embryonic skull are gradually replaced by bone. In large-brained mammals, and especially in humans, this process is not completed until after birth. Young babies have a soft spot, the fontanelle, on the top of the skull that allows the head to be compressed so it can pass through the birth canal.

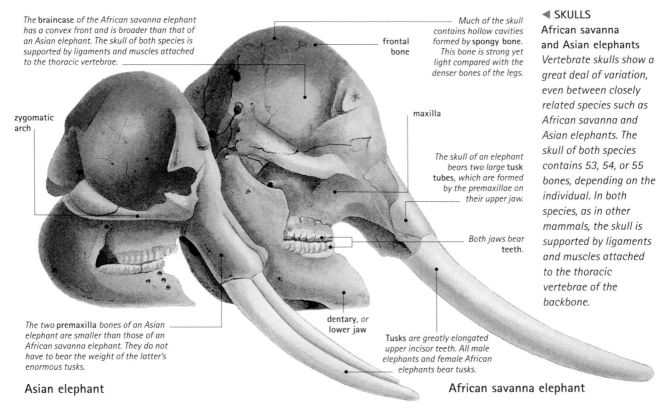

The braincase of the African savanna elephant has a convex front and is broader than that of an Asian elephant. The skull of both species is supported by ligaments and muscles attached to the thoracic vertebrae.

zygomatic arch

frontal bone

Much of the skull contains hollow cavities formed by spongy bone. This bone is strong yet light compared with the denser bones of the legs.

maxilla

The skull of an elephant bears two large tusk tubes, which are formed by the premaxillae on their upper jaw.

Both jaws bear teeth.

The two premaxilla bones of an Asian elephant are smaller than those of an African savanna elephant. They do not have to bear the weight of the latter's enormous tusks.

dentary, or lower jaw

Tusks are greatly elongated upper incisor teeth. All male elephants and female African elephants bear tusks.

Asian elephant

◄ SKULLS
African savanna and Asian elephants
Vertebrate skulls show a great deal of variation, even between closely related species such as African savanna and Asian elephants. The skull of both species contains 53, 54, or 55 bones, depending on the individual. In both species, as in other mammals, the skull is supported by ligaments and muscles attached to the thoracic vertebrae of the backbone.

African savanna elephant

Vertebrate axial skeleton

CONNECTIONS

COMPARE the neck bones of the *GIRAFFE* with those of a *WILDEBEEST*. Both have seven cervical vertebrae, despite the difference in neck length.

COMPARE the coccyx of the *HUMAN* with the pygostyle of the *PENGUIN* and the urostyle of the *BULLFROG*. All are evolved from the longer tails of their ancestors.

Vertebrates are by definition animals in which the notochord becomes gradually replaced by bony or cartilaginous units called vertebrae, which collectively form the backbone, or spine. However, in the simplest of vertebrates, this substitution is incomplete, and the notochord remains the main support structure throughout life. Indeed, in hagfish, the only evidence of any backbone is a row of tiny cartilaginous structures in the tail, forming the most rudimentary of vertebrae. In the lamprey, there are cartilaginous vertebrae along the full length of the notochord, but they are still relatively small.

The vertebrae of sharks and rays are much more substantial. Each one develops around the notochord, which is constricted into a shape resembling a string of beads. The part of the vertebra through which the notochord passes is called the centrum. In bony fish and tetrapods, the centrum is filled in with bone during embryonic development, so the notochord is constricted more and more and eventually broken into a series of cushions between the vertebrae. These cushions are the intervertebral disks.

Vertebrae

In addition to their supportive role, the vertebrae also protect the spinal cord. This vital tract of nerve runs along the back of the animal, through a tube created by a series of arches, one on the top of each vertebra. In some vertebrates there is a second set of arches—the hemal arches—below the centrum. The vertebrae may also bear a number of spine projections, or processes, which serve as attachment points for various muscles. The processes vary considerably in size and shape at different points along the backbone and between species, so a zoologist can determine the difference between, for example, the cervical (neck) vertebrae and the lumbar (back) vertebrae of an animal.

▶ THORACIC VERTEBRA
Whale
Whales have 14 thoracic vertebrae in the trunk region. These vertebrae do not permit much movement, but the ribs attached to them (one on each side) are mobile. Each vertebra is large, with broad surfaces and extensions called neural spines and transverse processes. The processes act as attachment points for muscles, and the neural spines act as levers, increasing the power that can be transmitted by the muscles to make the spine flex.

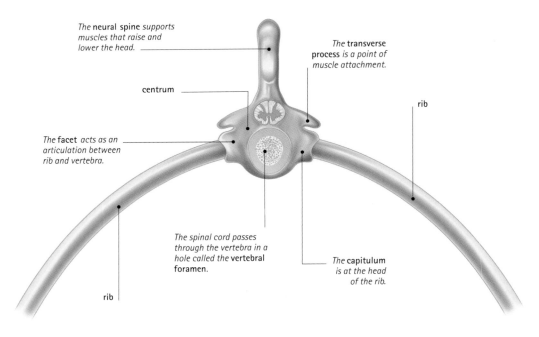

The **neural spine** supports muscles that raise and lower the head.

The **transverse process** is a point of muscle attachment.

centrum

rib

The **facet** acts as an articulation between rib and vertebra.

The spinal cord passes through the vertebra in a hole called the **vertebral foramen**.

The **capitulum** is at the head of the rib.

rib

The vertebrae at the top of the spine are called cervical vertebrae. The first of these, which articulates directly with the base of the skull, is the atlas. In mammals, the atlas has two depressions in the anterior (front) face nearest the skull; these fit snugly around two bumps in the base of the skull called occipital condyles. The bumps are two of the many skeletal features that set mammals apart from other vertebrates. In amphibians, the atlas is the only cervical vertebra. In "higher" vertebrates, the atlas articulates with the second cervical vertebra, or axis. Following the atlas and axis are a series of conventional cervical vertebrae, making a total of seven in most mammals but as many as 25 in some long-necked birds.

The largest, most complex vertebrae are usually those of the thoracic or upper back region. These often bear long dorsal processes for the attachment of large muscles associated with the neck and shoulders, and transverse processes that serve as attachment points for the ribs. The lumbar vertebrae of the lower back are somewhat similar in form, but with more modest processes. The vertebrae of the pelvic region are known as sacral vertebrae. Those of tetrapods are fused into a rigid structure, the sacrum, which supports the pelvis. In birds the sacrum is fused fully with the other pelvic bones, forming one large, very strong structure called the synsacrum.

Features for flight

Flight places a unique set of stresses and strains on the thoracic region. The sternum of flying birds is greatly enlarged and bears a pronounced keel to provide attachment for the enormous breast muscles required to power flight. Flightless birds do not have a large sternum. The enlarged coracoid bone of flying birds offers some support, and the ribs are reinforced with special cross supports called uncinate struts. These resist compression of the rib cage, allowing the birds to continue breathing deeply during flight.

The tail

A tail is another characteristic feature of vertebrates. The tail is a continuation of the vertebral column in which the vertebrae are smaller and more simple than elsewhere. The spinal cord does not continue into the tail, so

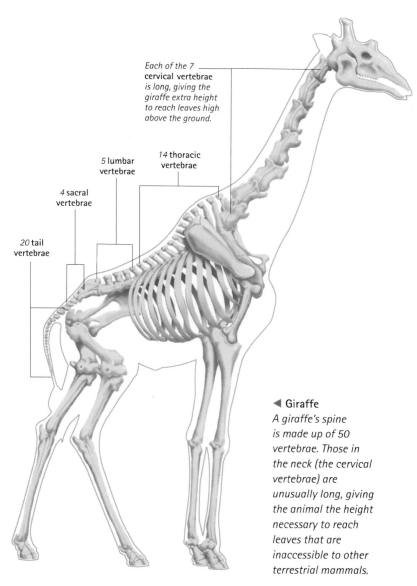

Each of the 7 **cervical vertebrae** *is long, giving the giraffe extra height to reach leaves high above the ground.*

5 lumbar vertebrae

14 thoracic vertebrae

4 sacral vertebrae

20 tail vertebrae

◀ **Giraffe**
A giraffe's spine is made up of 50 vertebrae. Those in the neck (the cervical vertebrae) are unusually long, giving the animal the height necessary to reach leaves that are inaccessible to other terrestrial mammals.

the caudal (tail) vertebrae lack a neural arch. In great apes, such as chimpanzees and humans, the tail is greatly reduced. It contains just four vertebrae, fused to form a bony bump—the coccyx—at the base of the spine. In birds, the tail is also reduced. The caudal vertebra are fused to form a structure called the pygostyle. This feature is an attachment point for the muscles that move the tail feathers, which play a vital role in controlling the aerodynamics of flight. In amphibians the tail vertebrae are fused to form the urostyle, a rod of bone that helps reinforce the pelvis.

Vertebrate appendicular skeleton

CONNECTIONS

COMPARE the lobe fin of a *COELACANTH* with the ray fin of a *TROUT*. It is easy to see which has more potential as a weight-bearing structure.

COMPARE the hind limbs of a *BULLFROG* with those of a *KANGAROO*. Both are long and powerful as an adaptation for a hopping mode of locomotion.

COMPARE the forelimb of the *PENGUIN* with that of a *DOLPHIN*. In both, the forelimbs function as flippers.

The term *appendicular skeleton* refers to the appendages (limbs and fins) and the girdles that attach them to the axial skeleton. In most vertebrates, there are two sets of paired appendages: pectoral appendages attached to a pectoral girdle at the front end of the animal and pelvic appendages attached to the pelvic girdle toward the back.

The girdles provide stability for the fins of fish, and in tetrapods they allow the weight of the body to be transferred through the limbs. The bones of the pectoral girdle are the scapulae or shoulder blades, the clavicles or collarbones, and the coracoid. In birds, the two clavicles are fused to make a single more rigid structure called the furcula or wishbone. The vertebrate pelvic girdle is made up of the ilium, ischium, and pubis, which may be separate or fused into one structure, the pelvis. The pelvic girdle of fish is a small, isolated structure embedded in soft tissue. That of tetrapods is much larger and is connected to the vertebral column in the region of the sacral vertebrae.

Fins

The paired fins of fish fall into two main skeletal types. By far the greater number of modern fish belong to the ray-finned group, called the Actinopterygii. Their pectoral and pelvic fins have no real skeletal support; instead, these fins have just a short stub of bones at the base. The fins themselves are webs of skin supported mainly by dermal rays, spokelike rods derived from the skin tissues. The more rigid fins of modern sharks and their relatives are supported by similar rods of cartilage. In contrast, the so called lobe-finned fish have robust, fleshy paired fins, supported by a much more substantial bony skeleton. There are many extinct groups of lobe-finned fish, including the Sacroptergii, which are probably the ancestors of tetrapods. Modern lobe-finned fish are now much less common and include lungfish and coelacanths.

The pentadactyl limb

The limbs of tetrapods (amphibians, reptiles, birds, and mammals) come in all shapes and

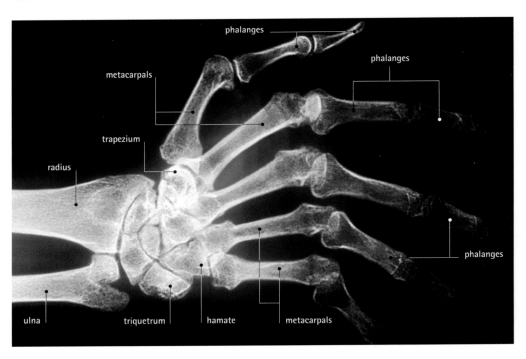

▶ *The five-fingered (pentadactyl) human hand has evolved from the pentadactyl plan of early amphibian ancestors. The hamate, the triquetrum, the trapezium, and several other small bones in the wrist are collectively called the carpals.*

Rhinoceros

The femur, tibia, and fibula are very robust and so able to support the heavy animal's weight. The rhino has an unguligrade stance: it walks on the tips of its toes.

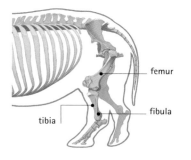

femur

fibula

tibia

Hyena

The femur, tibia, and fibula are long and slender, and these bones are the most vulnerable of a hyena's skeleton. The stance is digitigrade: walking on the toes.

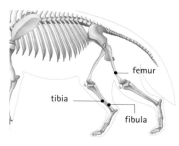

femur

tibia

fibula

▼ VESTIGIAL HIND LEG AND PELVIS
Green anaconda

Structures that no longer have any function are called vestigial. Green anacondas and other boas have vestigial hind legs and a pelvis, which indicate that their ancestors had limbs.

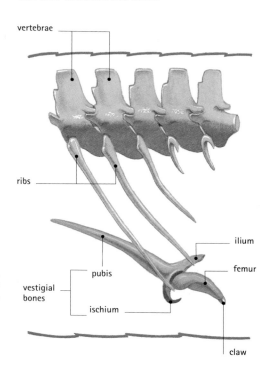

vertebrae

ribs

vestigial bones

pubis

ischium

ilium

femur

claw

Ostrich

The femur is short, but the lower leg bones—the tibiotarsus and tarsometatarsus—are very long and ideally suited for fast running.

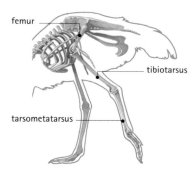

femur

tibiotarsus

tarsometatarsus

Kangaroo

The bones of the hind legs are robust, but whereas the femur is short, the tibia, fibula, and metatarsals are very long. This arrangement helps the kangaroo to leap.

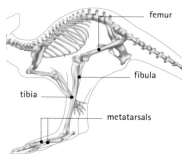

femur

fibula

tibia

metatarsals

sizes, from the flippers and flukes of seals and whales to the wings of bats, the slender hooved legs of horses and antelope, and the dexterous hands and feet of the great apes. However, from a skeletal point of view, all these structures are similar. They are nearly all based on the five-fingered pentadactyl limb inherited from an early amphibian ancestor. The pentadactyl limb contains a basic set of bones, which vary in size and structure to create the startling variety of forms described above.

Bats and birds

The versatility of the pentadactyl limb is demonstrated to great effect in the forelimbs of flying vertebrates. The five-fingered plan is obvious in the wing of a bat, which is supported by a set of greatly elongated hand bones. They support the membrane of the wing in much the same way as the ribs of an umbrella—

▲ HIND LIMB BONES

The remarkable diversity in the arrangement of vertebrates' hind bones reflects extremely varied lifestyles.

opening out to hold the wing taut and closing to fold the whole wing away. Bat wing bones are extremely thin, and this characteristic makes them light but not very strong.

The pentadactyl plan is less easily recognized in the wing of a bird, where the bones of the wrist, hand, and fingers are reduced in number and modified in size and shape. The wing bones of birds are larger than those of bats; but being hollow, they are also very light. Their role is mainly supportive. Unlike bats, whose skeleton dictates the precise shape of the wing, birds fine-tune the shape their wings by adjusting the angles of the feathers.

AMY-JANE BEER

FURTHER READING AND RESEARCH

Raven, Peter H., George B. Johnson, Susan R. Singer, and Jonathan B. Losos. 2004. *Biology*. McGraw-Hill Science: New York.

Index